CHIMPANZEES IN BIOMEDICAL AND BEHAVIORAL RESEARCH

ASSESSING THE NECESSITY

Committee on the Use of Chimpanzees in Biomedical
and Behavioral Research

Board on Health Sciences Policy
Institute of Medicine

Board on Life Sciences
Division on Earth and Life Studies

Bruce M. Altevogt, Diana E. Pankevich,
Marilee K. Shelton-Davenport, and Jeffrey P. Kahn, *Editors*

INSTITUTE OF MEDICINE *AND*
NATIONAL RESEARCH COUNCIL
OF THE NATIONAL ACADEMIES

THE NATIONAL ACADEMIES PRESS
Washington, D.C.
www.nap.edu

THE NATIONAL ACADEMIES PRESS • 500 Fifth Street, N.W. • Washington, DC 20001

This study was requested by Contract No. N01-OD-4-239 Task Order No. 248 between the National Academy of Sciences and the Department of Health and Human Services, National Institutes of Health. Any opinions, findings, conclusions, or recommendations expressed in this publication are those of the author(s) and do not necessarily reflect the view of the organizations or agencies that provided support for this project.

International Standard Book Number-13: 978-0-309-22039-2
International Standard Book Number-11: 0-309-22039-4

Additional copies of this report are available from The National Academies Press, 500 Fifth Street, N.W., Lockbox 285, Washington, DC 20055; (800) 624-6242 or (202) 334-3313 (in the Washington metropolitan area); Internet, http://www.nap.edu.

For more information about the Institute of Medicine, visit the IOM home page at: **www.iom.edu.**

Suggested citation: IOM (Institute of Medicine). 2011. *Chimpanzees in biomedical and behavioral research: Assessing the necessity.* Washington, DC: The National Academies Press.

THE NATIONAL ACADEMIES
Advisers to the Nation on Science, Engineering, and Medicine

The **National Academy of Sciences** is a private, nonprofit, self-perpetuating society of distinguished scholars engaged in scientific and engineering research, dedicated to the furtherance of science and technology and to their use for the general welfare. Upon the authority of the charter granted to it by the Congress in 1863, the Academy has a mandate that requires it to advise the federal government on scientific and technical matters. Dr. Ralph J. Cicerone is president of the National Academy of Sciences.

The **National Academy of Engineering** was established in 1964, under the charter of the National Academy of Sciences, as a parallel organization of outstanding engineers. It is autonomous in its administration and in the selection of its members, sharing with the National Academy of Sciences the responsibility for advising the federal government. The National Academy of Engineering also sponsors engineering programs aimed at meeting national needs, encourages education and research, and recognizes the superior achievements of engineers. Dr. Charles M. Vest is president of the National Academy of Engineering.

The **Institute of Medicine** was established in 1970 by the National Academy of Sciences to secure the services of eminent members of appropriate professions in the examination of policy matters pertaining to the health of the public. The Institute acts under the responsibility given to the National Academy of Sciences by its congressional charter to be an adviser to the federal government and, upon its own initiative, to identify issues of medical care, research, and education. Dr. Harvey V. Fineberg is president of the Institute of Medicine.

The **National Research Council** was organized by the National Academy of Sciences in 1916 to associate the broad community of science and technology with the Academy's purposes of furthering knowledge and advising the federal government. Functioning in accordance with general policies determined by the Academy, the Council has become the principal operating agency of both the National Academy of Sciences and the National Academy of Engineering in providing services to the government, the public, and the scientific and engineering communities. The Council is administered jointly by both Academies and the Institute of Medicine. Dr. Ralph J. Cicerone and Dr. Charles M. Vest are chair and vice chair, respectively, of the National Research Council.

www.national-academies.org

Reviewers

This report has been reviewed in draft form by individuals chosen for their diverse perspectives and technical expertise, in accordance with procedures approved by the National Research Council's Report Review Committee. The purpose of this independent review is to provide candid and critical comments that will assist the institution in making its published report as sound as possible and to ensure that the report meets institutional standards for objectivity, evidence, and responsiveness to the study charge. The review comments and draft manuscript remain confidential to protect the integrity of the deliberative process. We wish to thank the following individuals for their review of this report:

Stephen W. Barthold, University of California, Davis
Thomas M. Butler, Independent consultant
Alexander M. Capron, University of Southern California
Timothy Coetzee, National Multiple Sclerosis Society
Frans B. M. de Waal, Emory University
Jane Goodall, Jane Goodall Institute
Beatrice H. Hahn, University of Pennsylvania
Donald A. Henderson, Johns Hopkins University
William D. Hopkins, Agnes Scott College
Steven E. Hyman, Harvard University
Stanley M. Lemon, University of North Carolina at Chapel Hill
Alexander Ploss, The Rockefeller University
Arthur Weiss, University of California, San Francisco

Although the reviewers listed above have provided many constructive comments and suggestions, they were not asked to endorse the conclusions or recommendations, nor did they see the final draft of the report before its release. The review of this report was overseen by **Eli Y. Adashi,** Immediate Past Dean of Medicine & Biological Sciences, Brown University, and **Peter H. Raven,** President Emeritus, Missouri Botanical Garden. Appointed by the National Research Council and Institute of Medicine, they were responsible for making certain that an independent examination of this report was carried out in accordance with institutional procedures and that all review comments were carefully considered. Responsibility for the final content of this report rests entirely with the authoring committee and the institution.

Contents

Summary

At the request of National Institutes of Health (NIH), and in response to congressional inquiry, the Institute of Medicine (IOM) in collaboration with the National Research Council (NRC) convened an ad hoc committee to consider the necessity of the use of chimpanzees in NIH-funded research in support of the advancement of the public's health.

Specifically, the committee was asked to review the current use of chimpanzees for biomedical and behavioral research and:

- Explore contemporary and anticipated biomedical research questions to determine if chimpanzees are or will be necessary for research discoveries and to determine the safety and efficacy of new prevention or treatment strategies. If biomedical research questions are identified:

  o Describe the unique biological/immunological characteristics of the chimpanzee that make it the necessary animal model for use in the types of research.
  o Provide recommendations for any new or revised scientific parameters to guide how and when to use these animals for research.

- Explore contemporary and anticipated behavioral research questions to determine if chimpanzees are necessary for progress in understanding social, neurological, and behavioral factors that influence the development, prevention, or treatment of disease.

In addressing the task, the committee explored existing and anticipated alternatives to the use of chimpanzees in biomedical and behavioral research. The committee based its findings and recommendations on available scientific evidence, published literature, public testimony, submitted materials by stakeholders, and a commissioned paper, as well as its expert judgment.

To conduct this expert assessment and evaluate the necessity for chimpanzees in research to advance the public's health, the committee deliberated from May 2011 through November 2011. During this period, the committee held three 2-day meetings and several conference calls, including two public information-gathering sessions on May 26, 2011, and August 11-12, 2011. Each information-gathering session included testimony from individuals and organizations that both supported and opposed the continued use of chimpanzees. The committee also reviewed a number of background documents provided by stakeholder organizations and commissioned a paper, "Comparison of Immunity to Pathogens in Humans, Chimpanzees, and Macaques."

The committee identified a set of core principles and criteria that were used to assess the necessity of chimpanzees for research now or in the future.

Ethical Considerations

Neither the cost of using chimpanzees in research nor the ethical implications of that use were specifically in the committee's charge. Rather, the committee was asked for its advice on the scientific necessity of the chimpanzee model for biomedical and behavioral research. The committee agrees that cost should not be a consideration. However, the committee feels strongly that any assessment of the necessity for using chimpanzees as an animal model in research raises ethical issues, and any analysis of necessity must take these ethical issues into account. The committee's view is that the chimpanzee's genetic proximity to humans and the resulting biological and behavioral characteristics not only make it a uniquely valuable species for certain types of research, but also demand a greater justification for conducting research using this animal model.

Summary of Chimpanzee Research

The committee was asked, as part of its task, to review the current use of chimpanzees for biomedical and behavioral research. To assess the use of the chimpanzee as an animal model, the committee explored research supported by the NIH and other federally and privately funded research over the past 10 years.

The largest percentage of federally funded chimpanzee research has been supported by the NIH, with additional projects funded by other federal agencies, including the Food and Drug Administration (FDA), Centers for Disease Control and Prevention (CDC), and National Science Foundation. Of the 110 identified projects sponsored by the NIH between 2001 and 2010, 44 were for research on hepatitis; comparative genomics accounted for 13 projects; 11 projects were for neuroscience research; 9 projects were for AIDS/HIV studies; and 7 projects were for behavioral research. The remaining projects funded a limited number of studies in areas such as malaria and respiratory syncytial virus and projects supporting chimpanzee colonies.

Committee analysis of the use of chimpanzees in the private sector was hindered by the proprietary nature of the information. However, based on limited publications and public non-proprietary information, it is clear that the private sector is using the chimpanzee model, especially in areas of drug safety, efficacy, and pharmacokinetics. Although its use appears to be limited and decreasing over the 10 years examined by the committee, the chimpanzee model is being employed by industry in the development of antiviral drugs and vaccines for hepatitis B and C as well as in the development of monoclonal antibody therapeutics.

Principles Guiding the Use of Chimpanzees in Research

The task given to the committee by the NIH asked two questions about the need for chimpanzees in research: (1) Is biomedical research with chimpanzees "necessary for research discoveries and to determine the safety and efficacy of new prevention or treatment strategies?" and (2) Is behavioral research using chimpanzees "necessary for progress in understanding social, neurological, and behavioral factors that influence the development, prevention, or treatment of disease?" In responding to these questions, the committee concluded that the potential reasons for undertaking biomedical and behavioral research as well as the protocols used in each area are different enough to require different sets of criteria.

However, the committee developed both sets of criteria guided by the following three principles:

1. The knowledge gained must be necessary to advance the public's health;
2. There must be no other research model by which the knowledge could be obtained, and the research cannot be ethically performed on human subjects; and
3. The animals used in the proposed research must be maintained either in ethologically appropriate physical and social environments or in natural habitats.

These principles are the basis for the specific criteria that the committee established to assess current and future use of the chimpanzee in biomedical and behavioral research (see Recommendations 1 and 2).

Conclusions and Recommendations

The committee based the following conclusions and recommendations in large part on the advances that have been made by the scientific community using alternative models to the chimpanzee, such as studies using other non-human primates, genetically modified mice, in vitro systems, and in silico technologies as well as human clinical trials. Having reviewed and analyzed contemporary and anticipated biomedical and behavioral research, the committee concludes that:

- No uniform set of criteria is currently used to assess the necessity of the chimpanzee in NIH-funded biomedical and behavioral research.
- While the chimpanzee has been a valuable animal model in past research, most current use of chimpanzees for biomedical research is unnecessary, based on the criteria established by the committee, except potentially for two current research uses:

   o Development of future monoclonal antibody therapies will not require the chimpanzee, due to currently available technologies. However, there may be a limited number of monoclonal antibodies already in the developmental pipeline that may require the continued use of chimpanzees.

o The committee was evenly split and unable to reach consensus on the necessity of the chimpanzee for the development of a prophylactic hepatitis C virus (HCV) vaccine. Specifically, the committee could not reach agreement on whether a preclinical challenge study using the chimpanzee model was necessary and if or how much the chimpanzee model would accelerate or improve prophylactic HCV vaccine development.

- The present trajectory indicates a decreasing scientific need for chimpanzee studies due to the emergence of non-chimpanzee models and technologies.
- Development of non-chimpanzee models requires continued support by the NIH.
- A new, emerging, or reemerging disease or disorder may present challenges to treatment, prevention, and/or control that defy non-chimpanzee models and available technologies and therefore may require the future use of the chimpanzee.
- Comparative genomics research may be necessary for understanding human development, disease mechanisms, and susceptibility because of the genetic proximity of the chimpanzee to humans. It poses no risk to the chimpanzee when biological materials are derived from existing samples or minimal risk of pain and distress in instances where samples are collected from living animals.
- Chimpanzees may be necessary for obtaining otherwise unattainable insights to support understanding of social and behavioral factors that include the development, prevention, or treatment of disease.
- Application of the committee's criteria would provide a framework to assess scientific necessity to guide the future use of chimpanzees in biomedical, comparative genomics, and behavioral research.

Recommendation 1: The National Institutes of Health should limit the use of chimpanzees in biomedical research to those studies that meet the following three criteria:

1. There is no other suitable model available, such as in vitro, non-human in vivo, or other models, for the research in question;
2. The research in question cannot be performed ethically on human subjects; and
3. Forgoing the use of chimpanzees for the research in question will significantly slow or prevent important advancements to prevent, control, and/or treat life-threatening or debilitating conditions.

Animals used in the proposed research must be maintained either in ethologically appropriate physical and social environments or in natural habitats. Biomedical research using stored samples is exempt from these criteria.

Recommendation 2: The National Institutes of Health should limit the use of chimpanzees in comparative genomics and behavioral research to those studies that meet the following two criteria:

1. Studies provide otherwise unattainable insight into comparative genomics, normal and abnormal behavior, mental health, emotion, or cognition; and
2. All experiments are performed on acquiescent animals, using techniques that are minimally invasive, and in a manner that minimizes pain and distress.

Animals used in the proposed research must be maintained either in ethologically appropriate physical and social environments or in natural habitats. Comparative genomics and behavioral research using stored samples are exempt from these criteria.

The criteria set forth in the report are intended to guide not only current research policy, but also decisions regarding potential use of the chimpanzee model for future research. The committee acknowledges that imposing an outright and immediate prohibition of funding could cause unacceptable losses to research programs as well as have an impact on the animals. Therefore, although the committee was not asked to consider how its recommended policies should be implemented, it believes that the assessment of the necessity of the chimpanzee in all grant renewals and future research projects would be strengthened and the process made more credible by establishing an independent oversight committee that builds on the Interagency Animal Model Committee and uses the recommended criteria.

STUDY BACKGROUND
AND CONTEXT

The chimpanzee (*Pan troglodytes*) is a current animal model in biomedical and behavioral research supported by the U.S. government and industry. In fiscal year 2011, of the more than 94,000 active projects sponsored by the National Institutes of Health (NIH), only 53 used the chimpanzee (0.056 percent). However, members of the public, Congress, and some scientists question this use. They argue that research that has relied on chimpanzees could be accomplished using other models, methods, or technologies (Bailey, 2010a, 2010b; Bettauer, 2011) or that chimpanzees are not appropriate models for human disease research (Bailey, 2008; Physicians Committee for Responsible Medicine, 2011).

Ongoing biomedical and behavioral research on chimpanzees is largely conducted at four facilities: the Southwest National Primate Research Center, the New Iberia Research Center at the University of Louisiana-Lafayette, the Michale E. Keeling Center for Comparative Medicine and Research of the University of Texas MD Anderson Cancer Center, and the Yerkes National Primate Research Center at Emory University. Much of the research supported by the first three facilities is focused on proof-of-principle studies for hepatitis C vaccines and therapies, with a lesser amount of research devoted to assessing safety and efficacy of large molecules such as monoclonal antibodies (Watson, 2011). In addition, research supports studies on deriving chimpanzee cell lines, antibodies and other biological materials, as well as comparative genomics research. The Yerkes Center primarily sponsors studies pertaining to developmental and cognitive neuroscience, as well as aging-related comparative neurobiology (Yerkes National Primate Research Center, 2011). In addition to these four centers, the National Center for Research Resources (NCRR) also supports the Alamogordo Primate Facility (APF). Unlike the other facilities, Alamogordo is a research reserve facility that does not have an active chimpanzee research program; no invasive research is conducted on these chimpanzees while on the premises[1] (NCRR, 2011a). However, the animals may be used for cardiovascular disease and behavioral studies with data obtained during their annual physicals (Watson, 2011). If these chimpanzees are needed

---

[1]According to solicitation NHLBI-CSB-(RR)-SS-2011-264-KJM (HHS, 2011c), "the current agreements between the National Institutes of Health (NIH) and the U.S. Air Force (USAF) prescribe that no invasive research shall be conducted on chimpanzees currently held at the APF."

for other types of research, they are relocated to another facility and, once removed, cannot return to Alamogordo Primate Facility (HHS, 2011d).

As of May 2011, 937 chimpanzees, ranging in age from less than 1 year old to greater than 41, were available for biomedical and behavioral research (Tables 1 and 2). The U.S. government currently supports 436 of these animals at four NCRR-supported facilities; the remaining animals are privately owned and supported (HHS, 2011a). The NCRR at the NIH provides programmatic oversight of these facilities and ensures they comply with the Animal Welfare Act, and with policies concerning laboratory animal care and use. Within the NCRR, the Division of Comparative Medicine oversees the NIH Chimpanzee Management Program (ChiMP), which supports the long-term, cost-effective housing and maintenance of chimpanzee facilities (NCRR, 2011a).

In 1995, the NIH instituted a moratorium on the breeding of chimpanzees that they owned or supported (NCRR, 2011b). Soon after, the Chimpanzee Management Plan Working Group was created to periodically assess the need for chimpanzees in research and report its findings to NCRR's advisory body, the National Advisory Research Resources Council. This Working Group of non-government scientists and non-scientists analyzes relevant issues and drafts proposed position papers. In 2007, this Working Group issued a report[2] that "did not make a definitive recommendation as to whether the chimpanzee breeding moratorium should be continued,"[3] but the NIH National Advisory Research Resources Council extended the breeding moratorium indefinitely (Cohen, 2007b). Given the life expectancy of chimpanzees in captivity, it is estimated that by 2037 the federally funded chimpanzee research population will "largely cease to exist" in the United States (Cohen, 2007a; NCRR, 2007).

---

[2] Report of the Chimpanzee Management Plan Working Group—March 9, 2007 (NCRR, 2007).

[3] The 1997 National Research Council report, *Chimpanzees in Research: Strategies for their Ethical Care, Management, and Use* also recommended a 5-year breeding moratorium (NAS, 1997).

TABLE 1 Number of Chimpanzees Available in the United States for Research

	Total Number of Chimpanzees[a]	Number of Chimpanzees Supported by the NCRR, NIH[b]
Alamogordo Primate Facility	176	176
Michale E. Keeling Center for Comparative Medicine and Research	176	159
New Iberia Research Center	347	124
Southwest National Primate Research Center	153	153
Yerkes National Primate Research Center[c]	85	0
TOTAL	937	612

[a]Number of chimpanzees as of October 2011 (Abee, 2011c; Else, 2011; Lammey, 2011; Landry, 2011; Langford, 2011).

[b]Number of NIH-supported chimpanzees current as of April 15, 2011 (HHS, 2011a).

[c]The Yerkes National Primate Research Center does not use any core funds from the NCRR to support the costs for maintaining humane care and welfare of chimpanzees.

TABLE 2 Ages of Chimpanzees Available in the United States for Research[a,b]

	< 10	10 to 20	21 to 30	31 to 40	41+
Alamogordo Primate Facility	0	24	99	40	13
Michale E. Keeling Center for Comparative Medicine and Research	0	53	67	27	29
New Iberia Research Center	100	134	84	6	23
Southwest National Primate Research Center	4	61	69	13	5
Yerkes National Primate Research Center[c]	1	29	30	12	13
TOTAL	105	301	349	98	83

[a]Ages of chimpanzees as of October 2011 (Abee, 2011c; Else, 2011; Lammey, 2011; Landry, 2011; Langford, 2011).

[b]The committee was unable to match the age of each chimpanzee with the funding source. Numbers represent a mix of federal and other sources of funding.

[c]The Yerkes National Primate Research Center does not use any core funds from the NCRR to support the costs for maintaining humane care and welfare of chimpanzees.

Origin of Study and Committee Statement of Task

The formation of the present committee activity and subsequent report was precipitated by events that took place in 2010, when the NIH announced its decision to transfer the chimpanzees located at the Alamogordo Primate Facility to the Southwest National Primate Research Center, where they would be consolidated with the chimpanzee colony that was already maintained and available for research (HHS, 2011b, 2011d). As the NIH's 10-year contract with Charles River Laboratories to manage the Alamogordo Primate Facility neared its completion, the NIH stated that consolidating the chimpanzees into a single colony at the Southwest National Primate Research Center facility would save $2 million a year and make the animals available for future research (HHS, 2011a; Korte, 2010). This decision stirred controversy. Animal rights activists and primate experts objected to returning the Alamogordo chimpanzees to a location where research is allowed, advocating instead for their permanent retirement (The Humane Society of the United States, 2010). Then–New Mexico Governor Bill Richardson also objected to closing the facility, which employs about 35 people (Korte, 2010). He asked the NIH to reverse its plans and requested that the U.S. Department of Agriculture (USDA) formally evaluate the way in which relocation plans were made. Governor Richardson requested that the Alamogordo Primate Facility be converted to an official sanctuary[4] or be operated by local universities for non-invasive behavioral research (Ledford, 2010).

In December 2010, amid increasing attention to the issue,[5] U.S. Senators Jeff Bingaman (D-NM), Tom Harkin (D-IA), and Tom Udall (D-NM) requested the National Academies conduct an in-depth analysis of the current and future need for chimpanzee use in biomedical research, an analysis they anticipated would consider the "great progress the science

---

[4]The U.S. Chimpanzee Health Improvement, Maintenance, and Protection Act of 2000 (106th Cong., 2nd sess.) required sanctuaries to house chimpanzees no longer needed for medical research.

[5]While not directly related to this study, it is of historical interest that bills were introduced in the U.S. Congress in 2008, 2009, 2010, and 2011 to ban research using chimpanzees and other great apes. Legislation included the Great Ape Protection Act of 2008, 110th Cong., 2d sess.; Great Ape Protection Act of 2009, 111th Cong., 1st sess.; Great Ape Protection Act of 2010, 111th Cong., 2d sess.; and Great Ape Protection and Cost Savings Act of 2011, 112th Cong., 1st sess. To date, the bills have not been adopted into law; however, activities related to the proposed legislation have also contributed to the national discussion about the necessity of chimpanzees for research.

community has made in research techniques" and "allow our nation's research institutions to make long-range decisions about the merits of continued invasive research using chimpanzees." In January 2011, the NIH announced it would suspend transfer of the Alamogordo colony and that it had tasked the Institute of Medicine (IOM) to study this issue (HHS, 2011b). Upon completion of this study, the NIH will revisit its decision regarding the Alamogordo colony.

In response to the request from the NIH, the IOM, in collaboration with the National Research Council, assembled the Committee on the Use of Chimpanzees in Biomedical and Behavioral Research to conduct a study and issue a report on the use of chimpanzees in NIH-funded research that is needed for the advancement of the public's health. The committee's statement of task is in Box 1.

BOX 1
Statement of Task

In response to a request from the National Institutes of Health (NIH), the Institute of Medicine, in collaboration with the National Research Council, will assemble an ad hoc expert committee that will conduct a study and issue a letter report on the use of chimpanzees in NIH-funded research that is needed for the advancement of the public's health. The primary focus will be animals owned by the National Institutes of Health, but will also include consideration of privately owned animals that are currently financially supported by the NIH.

Specifically, the committee will review the current use of chimpanzees for biomedical and behavioral research and:

- Explore contemporary and anticipated biomedical research questions to determine if chimpanzees are or will be necessary for research discoveries and to determine the safety and efficacy of new prevention or treatment strategies. If biomedical research questions are identified:

  o Describe the unique biological/immunological characteristics of the chimpanzee that make it the necessary animal model for use in the types of research.
  o Provide recommendations for any new or revised scientific parameters to guide how and when to use these animals for research.

- Explore contemporary and anticipated behavioral research questions to determine if chimpanzees are necessary for progress in understanding social, neurological, and behavioral factors that influence the development, prevention, or treatment of disease.

In addressing the task, the committee will explore contemporary and anticipated future alternatives to the use of chimpanzees in biomedical and behav-

---

ioral research that will be needed for the advancement of the public's health. The committee will base its findings and recommendations on currently available protocols, published literature, and scientific evidence, as well as its expert judgment.

Ethical Considerations

This report is based on the committee's evaluation of the ongoing chimpanzee research and its expert judgment and assessment of the needs for chimpanzee research. Neither the cost of using chimpanzees in research nor the ethical implications of that use were specifically in the committee's charge. Rather, the committee was asked for its advice on the scientific necessity of the chimpanzee as a human model for biomedical and behavioral research. The committee agrees that cost should not be a consideration. However, it recognizes that any assessment of the necessity for using chimpanzees as an animal model in research raises ethical issues, and any analysis must take these ethical issues into account. The committee's view is that the chimpanzee's genetic proximity to humans and the resulting biological and behavioral characteristics not only make it a uniquely valuable species for certain types of research, but also demand a greater justification for their use in research than is the case with other animals. Reports over many decades have established the principles and guidelines dictating that animal subjects must be used in studies only where the risk to the health and welfare of humans is too great (European Union, 2010; NAS, 2010; Parliament of the United Kingdom, 1987). Chimpanzees share biological, physiological, behavioral, and social characteristics with humans, and these commonalities may make chimpanzees a unique model for use in research. However, this relatedness—the closeness of chimpanzees to humans biologically and physiologically—is also the source of ethical concerns that are not as prominent when considering the use of other species in research. This is consistent with the 2010 European Union Directive, which notes that ethical issues are raised by the genetic proximity to human beings (European Union, 2010).

In simplest terms and following the committee's focus on necessity, the research use of animals that are so closely related to humans must offer insights not possible when using other animal models. In addition, the research must be of sufficient scientific or health value to offset these moral costs. There are many ethical approaches to analyze and either justify or proscribe the use of animals in research, and the committee was neither tasked nor appropriately composed to evaluate and reach consensus on a particular approach or to apply it to research on chimpanzees. However, in animal research policy, utilitarian justifications form part of the rationale for continued research in animals; that is, animals are subjected to risk for the benefit of humans, and justification relies on assessments that the benefits gained from research on animals are sufficient to outweigh the harms caused in the process. Purely utilitarian justifications are tempered in animal research through policy requirements for humane treatment and the use of appropriate species and minimal number of animals. Furthermore, imposing requirements for justifying the use of higher species is an implicit recognition that the use of higher animals comes at higher moral costs. Thus, the use of chimpanzees should face the most stringent requirements for justification, and constraints that acknowledge the characteristics that make chimpanzees unique among animal research subjects. For the committee, this ethical context is reflected in its assessment of when, if ever, the use of chimpanzees in biomedical research is necessary.

METHODS AND ORGANIZATION OF THE REPORT

To conduct this expert assessment and evaluate the need for chimpanzees in research to advance the public's health, the committee deliberated from May through November 2011. During this time, the committee held three 2-day meetings and several conference calls, including two public information-gathering sessions on May 26, 2011, and August 11-12, 2011 (see Appendix C for meeting agendas). Each information-gathering session included testimony from individuals and organizations that both supported and opposed the continued use of chimpanzees. The objectives of the information-gathering sessions were to:

- Obtain background data on the current use of chimpanzees in biomedical and behavioral research;
- Explore potential alternative models to chimpanzees; and
- Seek public comment about the scientific need for chimpanzees in biomedical and behavioral research.

In addition, during the course of the study the committee solicited and received over 5,700 comments via the Internet.

The committee examined the current availability of chimpanzees and use of the chimpanzee as an animal model. The committee also reviewed the use of chimpanzees in the peer-reviewed scientific literature, as described later in the section titled "Summary of Chimpanzee Research." In addition, it reviewed NIH projects that supported chimpanzee research from 2001 to 2010. The committee reviewed a number of background documents provided by stakeholder organizations. The committee also commissioned a paper titled "Comparison of Immunity to Pathogens in Humans, Chimpanzees, and Macaques" (see Appendix B).

The committee completed its task by identifying a set of core principles to guide current and future use of the chimpanzee, and based on these principles derived a set of criteria used to assess whether chimpanzees are necessary for research now or in the future.

INTERNATIONAL POLICIES GUIDING CHIMPANZEE USE

Many countries have legislation banning the use of great apes, and therefore chimpanzees.[6] Legal action may have been deemed unnecessary in countries where chimpanzee biomedical and behavioral research no longer occurs. The most recent legislative action around great ape use took place within the European Union (EU), with its 27 member states. In November 2010, following an eight-year political process, the EU adopted Directive 2010/63 outlining the protection of animals used for research purposes (European Union, 2010). This directive bans the use of great apes in research (Article 8), except for a specific safeguard clause that is described below (Article 55). Limitation of the ban to great apes, but not other non-human primates, and inclusion of the safeguard clause were based on political compromise that occurred over several years. Factors in the development of this compromise may have included

- No research using chimpanzees has been conducted at an EU facility since 1999 (European Parliament, 2007; Vogel, 2001);
- The last facility to house chimpanzees stopped all research in 2004 (BPRC, 2011);

---

[6]As will be discussed later in the "Summary of Chimpanzee Research" section, the committee did find that investigators from countries outside the United States have supported limited use of chimpanzees in the United States.

- Support by the European Commissions' Scientific Committee on Health and Environmental Risks (SCHER) for the continued use of non-human primates (NHPs) (Bateson, 2011; SCHER, 2009); and
- Recognition of the claims by the research community that the direction of new research is by definition unpredictable, as are the development of epidemics and emergence of new diseases.

The safeguard clause states that the use of great apes is permitted only for the purposes of research aimed at the preservation of those species or where action in relation to a potentially life-threatening, debilitating condition endangering human beings is warranted, and no other species or alternative method would suffice in order to achieve the aims of the procedure. While this clause was already in place in the previous version of the directive (European Communities and Office for Official Publications, 1986; Hartung, 2010), further details in the new directive (European Union, 2010) stipulate that in order for a member state to authorize a study involving great apes the member state must obtain approval from the European Commission in consultation with a relevant Committee (European Communities and Office for Official Publications, 1986) and (European Union, 2010). At the time of this report, Directive 2010/63 is still to be implemented in all European Union member states.

A number of countries, including EU member states, have specific laws or regulations involving the use of great apes and in some cases other NHPs (Table 3). The committee was unable to find any official policies guiding the use of chimpanzees in biomedical and behavioral research in other countries with large research investments, such as China and India, or to determine whether these countries maintain research populations of chimpanzees.

TABLE 3 International Policies on the Use of Great Apes in Scientific Research

Country or Entity (Year)	Policy or Statement
Australia (2003)	Restricts research and stipulates that "great apes may only be used for scientific purposes if the following conditions are met: Resources, including staff and house, are available to ensure high standards of care for the animals; the use would potentially benefit the individual animal and the species to which the animal belongs; the potential benefits of the scientific knowledge gained will outweigh harm to the animal" (Australian Government National Health and Medical Research Council, 2003).
Netherlands (2003)	The principal law on animal experimentation was amended with the insertion of a new Sec. 10e, which prohibits experimentation on chimpanzees, bonobos, orangutans, and gorillas. An exception was made in the case of experiments commenced before January 1, 2003, in which chimpanzees were used with a view to developing a vaccine against hepatitis C (WHO, 2003).
New Zealand (1999)	The Animal Welfare Act stipulates that the Director-General must not give approval unless he or she is satisfied that the use of the non-human hominid in that research, testing, or teaching either (1) it is in the best interests of the non-human hominid; or (2) it is in the interests of the species to which the non-human hominid belongs and that the benefits to be derived from the use of the non-human hominid in the research, testing, or teaching are not outweighed by the likely harm to the non-human hominid (Animal Welfare Act 1999 [New Zealand], 1999).

Country or Entity (Year)	Policy or Statement
Spain (2008)	The Commission on Environment, Agriculture and Fishing submitted a proposal to the Spanish Parliament to approve a resolution urging the country to comply with the Great Apes Project, founded in 1993, which argues that non-human great primates—chimpanzees, gorillas, orangutans and bonobos, should have the right to life, the protection of individual liberty, and the prohibition of torture (Congress of Spain, 2008).
United Kingdom (1997)	In November 1997, the government issued a supplementary note to its response to an interim report in which it published a policy statement on the use of animals in scientific procedures. It promised: the use of great apes in scientific procedures would not be allowed. While such animals have never been used under the 1986 Act, the government decided that it would be unethical to use such animals for research purposes due to their cognitive and behavioral characteristics and qualities. In the Home Office "News Release" accompanying the publication of the Interim Report, Lord Williams is quoted as follows: *"Although these proposed bans cannot be statutory under current legislation, I do not foresee any circumstances in which the Home Office would issue licenses in such cases"* (Reynolds and CEECE, 2001; Secretary of State for the Home Department and Parliament of the United Kingdom, 1998).

SUMMARY OF CHIMPANZEE RESEARCH

The committee was asked, as part of its task, to review the current use of chimpanzees for biomedical and behavioral research. To assess the use of the chimpanzee as an animal model, the committee explored research supported by the NIH and other federally and privately funded research over the past 5 years, and where possible, 10 years. A summary of this analysis is presented in the following section.

Analysis of Federally Supported Research

The largest percentage of federally funded chimpanzee research over the past 10 years has been supported by the NIH, with additional projects funded by other federal agencies, including the Food and Drug Administration (FDA), Centers for Disease Control and Prevention (CDC), and National Science Foundation (NSF).

NIH-Supported Research

To explore NIH-supported research, the committee used the Research Portfolio Reporting Tools Expenditures and Results (RePORTER) system to search for projects that included the terms "chimpanzee(s)" or "Pan troglodyte(s)." The search, conducted on July 6, 2011, was refined to exclude projects that were found to not use chimpanzees. Finally, the projects were categorized. From 2001 to 2010, the NIH funded 110 projects that used chimpanzees, chimpanzee genomic sequences, or other chimpanzee-derived compounds (Table 4). Hepatitis research,[7] the largest category with 44 projects, has included projects that range from molecular studies of the virus to immune responses in chimpanzees chronically infected with hepatitis C. In addition, studies have examined the pathogenesis of acute and chronic liver disease following infection. Comparative genomics studies included analysis of human and chimpanzee polymorphism rates. Some of the 11 neuroscience research projects focused on studies of neurodevelopment, while behavioral research studies examined task engagement and sociocommunicative development.[8]

---

[7]The term "hepatitis" is inclusive of all types of hepatitis, including A, B, C, D, and E.

[8]Behavioral research studies may also fall under additional categories, such as neuroscience.

Additional research areas included acquired immune deficiency syndrome (AIDS)/human immunodeficiency virus (HIV), malaria, and immunology. Of the remaining 22 projects, a portion was for research on a variety of topics, including studies of respiratory syncytial virus (RSV) and vaccines against anthrax toxin, while the remaining group of projects supported chimpanzee colonies, including the care and maintenance of the animals. Because each project varied in the number of years of funding, a breakdown of the number of research projects ongoing in each year in each disease category was performed (Figure 1). The number of annually funded NIH projects varied from 38 projects in 2002 to 52 in 2007.

TABLE 4 Number of Projects and Types of Funding per Disease Area: 2001-2010

| | Projects | Types of Funding | | | | |
		R^1	P^2	N^3	Z^4	U^5
Hepatitis	44	14	0	0	25	5
Comparative genomics	13	11	1	0	0	1
Neuroscience	11	7	1	0	3	0
AIDS/HIV	9	8	0	0	1	0
Behavioral	7	7	0	0	0	0
Malaria	2	2	0	0	0	0
Immunology	2	2	0	0	0	0
Other	11	2	0	0	8	1
Colony maintenance	11	2	1	3	0	5
TOTAL	110	55	3	3	37	12

[1]Research project grants.
[2]Program project/research center grants.
[3]Research contracts.
[4]Intramural grants.
[5]Cooperative agreements.

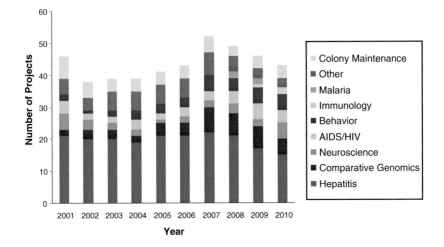

FIGURE 1 Chimpanzee research supported by the NIH: 2001-2010.

Other Federally Supported Research

Over the past decade, the FDA has funded a number of studies using chimpanzees, including the funding of the Laboratory of Hepatitis Research. The research supported by the FDA has focused on understanding the immunobiology and pathogenesis of hepatitis C virus and studying the safety of vaccines under development.

Other government agencies, including the NSF and CDC, have also funded chimpanzee research in the past 10 years, although to a significantly smaller degree than the NIH. During the past 3 years, the NSF has funded nine such studies, ranging from wild female chimpanzee emigration patterns to morphometric analysis of specific neocortical brain regions (NSF, 2011). Overall, the NSF has funded studies that include the use of both captive and wild chimpanzees, imaging data, and chimpanzee genomic information. While the CDC no longer funds chimpanzee research, previous research has included hepatitis vaccine development. Beyond these agencies, the committee did not find any evidence of current chimpanzee work funded by other federal agencies, including the Department of Defense.

Analysis of Private-Sector Supported Research

Animal models are used throughout discovery, development, preclinical testing, and production phases of new medicines and vaccines. Pharmaceutical and biotechnology companies use animals in the research and development of candidate compounds. In addition, regulatory agencies require that all new prescription drugs and biologics be subjected to thorough efficacy and safety testing prior to licensing. These requirements are in place to not only prevent potentially dangerous products from reaching human clinical trials and eventually the market, but also to ensure that only effective medications reach patients. In this context, many pharmaceutical companies state that NHPs are used when no other acceptable alternative exists and that the usual goal of using NHPs is to evaluate efficacy and safety as a final step prior to testing in humans. Several pharmaceutical companies no longer use chimpanzees, including GlaxoSmithKline, which has an official published policy indicating it has voluntarily ended the use of great apes, including chimpanzees, in research and will no longer initiate or fund studies (GlaxoSmithKline, 2011).

Committee analysis of the use of chimpanzees in the private sector was hindered by the proprietary nature of the information. However, based on limited publications and public non-proprietary information, it is clear that the private sector is accessing both the whole-animal model as well as stored biological samples (Carroll et al., 2009; Olsen et al., 2011). In addition, from data provided by the four NCRR-supported centers, the committee learned that from 2006 to 2010, 144 chimpanzees were used for efficacy, safety, and pharmacokinetic (PK) studies, suggesting that chimpanzees have been a part of the process of drug and/or vaccine development. These data do not make clear, however, which of these studies were funded by private companies and which, if any, were funded by the federal government. In addition, between 2005 and 2010, more than 300 requests for biological samples have come from individuals or groups with private funding, but, again, it was not possible to determine what percentage was funded by industry (Abee, 2011b; Langford, 2011; Rowell, 2011).

Use of chimpanzees in the United States is not limited to U.S.-based investigators, agencies, or companies. Between 2005 and 2010, 27 studies were funded by either non-U.S.-based companies or non-U.S.-based academic investigators (Watson, 2011). The majority of these studies were for hepatitis C therapy or vaccine development, with a few additional studies on monoclonal antibody efficacy and immunogenicity. Eight studies were funded by companies/investigators from Italy, followed by Japan and Denmark (five studies each). In addition, companies from Belgium, Spain, and France funded one study each. The committee hypothesizes that, among other reasons, foreign companies are using U.S. resources because of the EU ban on great ape research, the lack of research facilities in their respective countries capable of supporting chimpanzee research and, for industry, regulatory requirements both in the United States and abroad (Box 2).

While the committee was able to determine that both U.S.- and non-U.S.-based companies conduct limited chimpanzee research in the United States, it was not able to determine if companies independently house chimpanzees, how often the animals are used, and what compounds, if any, currently on the market or in human clinical trials were tested using this model.

BOX 2
Regulatory Requirements

Food and Drug Administration

The FDA regulatory policies regarding the approval of new drugs, vaccines, and other biological products do not specifically refer to chimpanzees. The FDA does provide guidance that safety and toxicology studies must be completed using the most appropriate, or relevant, species prior to preclinical testing. The FDA relies on the sponsor to select the species and demonstrate the usefulness of the model while encouraging dialogue between sponsors and the agency regarding the type of animal models considered for testing. While there are no official policies about the content of these dialogues, the committee was able to learn about internal, unwritten practices of different branches of the FDA. The Center for Drug Evaluation and Research (CDER) does not ask for chimpanzee data and specifically discourages the use of chimpanzees when approached by sponsors. This decision is based on, in part, the availability of other methods for developing the required data, including the use of transgenic and chimeric animals, surrogate antibodies, and the minimal anticipated biological effect level approach. CDER, however, does not turn away applications that contain chimpanzee data, including seven applications in the past 5 years. Like CDER, the Center for Biologics Evaluation and Research (CBER) does not have a specific policy on the use of chimpanzees and does not require their use, if the sponsor is able to demonstrate the relevance or appropriateness of a different animal model. However, in contrast to CDER, CBER does not actively discourage the use of chimpanzees, in particular for use in vaccine development to prove effectiveness or demonstrate safety.

The Special Case of the FDA's Animal Rule

In some selected circumstances, when it is not possible to conduct human studies, the FDA can grant marketing approval based the Animal Rule (FDA, 2011a, 2011b). The Animal Rule states that approval would require adequate and well-controlled animal studies whose results show that the drug or biologic is reasonably likely to produce clinical benefit in humans (CDER and CBER, 2009).

European Medicines Agency

European Union (EU) regulatory requirements related to marketing authorization of medical products do not specifically refer to chimpanzees, although there is some guidance on the use of the most sensitive and relevant species (EMEA, 2008; 2011a). Within the European Medicines Agency (EMEA), the Committee for Medicinal Products for Human Use (CHMP) is responsible for determining whether or not medicines meet quality, safety, and efficacy requirements (CHMP, 2011). In preclinical safety evaluation guidance, the CHMP defines a relevant species as "one in which the test material is pharmacologically active due to the expression of the receptor or an epitope (in the case of monoclonal antibodies)" (EMEA, 2011a). Additionally, the CHMP recommends that safety evaluation programs should include the use of two relevant species, although one species may be sufficient if justification is provided. The EMEA does not require or recommend the use of chimpanzees for product approval. However, should a marketing authorization application contain results from chimpanzee studies, this does not disqualify the product or data. Between 2004 and 2010, the EMEA has authorized nine products based, in part, on chimpanzee data (European record assessment reports). No marketing ban on medicines or vaccines developed using chimpanzees was provided for in current legislation. Directive 2010/63, which makes the ban of great apes more explicit, does not change anything in EMEA practice as there was no specific requirement for the use of chimpanzees in place before the revision of the previous directive.

Criteria That Guide the Current Use of Chimpanzees

Each chimpanzee research center has individual, but similar, processes by which a researcher has resource requests evaluated (Abee et al., 2011). At each center an ad hoc committee, composed of researchers, veterinarians, behavioral biologists, and other experts, reviews each request using a unique set of questions. These questions are designed to evaluate the study rationale, determine if the chimpanzee is needed, and then assess how many animals are required. The dialogue continues until either it is determined the chimpanzee is no longer required or every member of the advisory committee is convinced that the study will be conducted appropriately and that all the preliminary studies have been completed.

In addition to the review performed by the Chimpanzee Research Centers, additional reviews occur prior to the start of any chimpanzee study. For all projects, the investigator's institutional animal care and use committee must approve the study protocol. In addition, the NIH Inter-agency Animal Model Committee must determine that the chimpanzee is the appropriate model for any project approved by a Chimpanzee Research Center that will use an NIH-owned chimpanzee (Bennett et al., 1995; DHS, 2007). However, as is the case for the reviews performed by the Chimpanzee Research Centers, the Interagency Animal Model Committee does not evaluate protocols against a uniform set of criteria.

Finding

There are currently no uniform set of criteria used to assess the necessity of the chimpanzee in NIH-funded biomedical and behavioral research.

PRINCIPLES GUIDING THE USE OF CHIMPANZEES IN RESEARCH

The task given to the committee by the NIH asked two questions about the need for chimpanzees in research: (1) Is biomedical research using chimpanzees "necessary for research discoveries and to determine the safety and efficacy of new prevention or treatment strategies?" and (2) Is behavioral research with chimpanzees "necessary for progress in understanding social, neurological, and behavioral factors that influence the development, prevention, or treatment of disease?" In responding to these questions, the committee concluded that the potential reasons for undertaking biomedical and behavioral research as well as the protocols used in each area are different enough to require different sets of criteria. However, the committee developed both sets of criteria guided by the following three principles:

1. The knowledge gained must be necessary to advance the public's health;

2. There must be no other research model by which the knowledge could be obtained, and the research cannot be ethically performed on human subjects; and
3. The animals used in the proposed research must be maintained either in ethologically appropriate physical and social environments or in natural habitats.

Ethologically Appropriate Physical and Social Environments

Chimpanzee research should be permitted only on animals maintained in an ethologically appropriate physical and social environment or in natural habitats. Chimpanzees live in complex social groups characterized by considerable interindividual cooperation, altruism, deception, and cultural transmission of learned behavior (including tool use). Furthermore, laboratory research has demonstrated that chimpanzees can master the rudiments of symbolic language and numericity, that they have the capacity for empathy and self-recognition, and that they have the human-like ability to attribute mental states to themselves and others (known as the "theory of mind"). Finally, in appropriate circumstances, chimpanzees display grief and signs of depression that are reminiscent of human responses to similar situations. It is generally accepted that all species, including our own, experience a chronic stress response (comprising behavioral as well as physiological signs) when deprived of usual habitats, which for chimpanzees includes the presence of conspecifics and sufficient space and environmental complexity to exhibit species-typical behavior. Therefore, to perform rigorous (replicable and reliable) biomedical and behavioral research, it is critical to minimize potential sources of stress on the chimpanzee. This can be achieved primarily by maintaining animals on protocols either in their natural habitats, or by consistently maintaining with conspecifics in planned, ethologically appropriate physical and social environments in facilities accredited by the Association for Assessment and Accreditation of Laboratory Animal Care International (AZA Ape TAG, 2010; Council of Europe, 2006; NRC, 1997, 2010). Examples of appropriate physical and social environments currently accredited by the Association for Assessment and Accreditation of Laboratory Animal Care International include primadomes or corrals with environmental enrichment, outdoor caging with access to shelter, and indoor caging.

The committee recognizes exceptions to this criterion may be warranted. For example, as a result of previously approved protocols, there are currently a few long-term research projects in which the living conditions and the relationships with humans have been idiosyncratic and integral to the protocols (e.g., studies where a chimpanzee is being taught a symbolic language and lives and/or intensely interacts with a small number of researchers). In addition, current health and prior infectious exposures might prevent social housing for particular animals in potential experiments that may need to be performed in biosafety level (BSL) 3 or 4 facilities. Therefore, while the committee encourages that animals be maintained in planned, ethologically appropriate physical and social settings or natural habitats, existing protocols should be judged on a case-by-case basis, and changes made should impose minimal physiological and psychological harm to the animals and disruption to their existing relationships with people. All future studies should conform to the need for ethologically appropriate housing.

Criteria to Assess the Necessity of the Chimpanzee for Biomedical Research

As previously discussed, the chimpanzee raises unique considerations due to the ethical issues that arise as a result of the chimpanzee's genetic proximity to human beings. Therefore, based on the principles previously defined, the committee developed the following criteria to guide its assessment of NIH-funded biomedical research using the chimpanzee:

1. There is no other suitable model available, such as in vitro, non-human in vivo, or other models, for the research in question;
2. The research in question cannot be performed ethically on human subjects; and
3. Chimpanzees are necessary to accelerate prevention, control, and/or treatment of potentially life-threatening or debilitating conditions.

Specific and full scientific justification for use of the chimpanzee must meet all three of the above criteria. Assessment of which uses meet these criteria should be done prospectively on a study-by-study basis. It is important that justification is substantiated and provides adequate evidence; statements such as the following would not be acceptable to the committee:

- "The chimpanzee is immunologically, physiologically, anatomically, and/or metabolically similar to human beings." This statement is too broad.
- "Chimpanzees have previously been used in safety studies for this class of drug." This statement is not specific as to the science driving the decision.

It is important to note that the committee focused its task on the type of research supported by the NIH. The committee acknowledges that biomedical research aimed at the preservation and welfare of the chimpanzee species may also necessitate use of the chimpanzee, but this research is not be supported by the NIH unless it has direct application towards advancing human health and so on its own is outside the committee's task.

Assessing Suitability of Available In Vitro or Non-Human In Vivo Models

Continued advances over the past decade in imaging, genetics, in vitro, and in silico models, and sophisticated rodent disease models have provided scientists with more tools that could be used in place of the chimpanzee. Federal regulations require that animals selected for a protocol should be of an appropriate species and quality and that the minimum number required to obtain valid results should be used (U.S. Office of Laboratory Animal Welfare, 2002). Methods such as mathematical models, computer simulation, and in vitro biological systems should also be considered before chimpanzees are considered for research.

When assessing the necessity of the chimpanzee as a model, a more stringent process of eliminating ("deselecting") models of species less closely related to human beings should be required, similar to the process adopted by many countries in Europe (European Union, 2010). For example, in the United Kingdom, Section 5 of the Animals Scientific Procedures Act states that the Secretary of State may not authorize any procedures where an alternative exists (Parliament of the United Kingdom, 1987). The rationale for selection of the chimpanzee as the necessary model must be supported by facts and data (Box 3). The process must be rigorous and principles for deselection must be clearly defined and consistent across institutions.

BOX 3
Deselection Criteria

The following are specific examples of deselection criteria that the committee used to assess the suitability of available in vitro or non-human in vivo models.

In Vitro Culture System
In vitro models must be deselected if specific data required can only be obtained through the use of in vivo models.

In Vivo Models
Other species–such as monkeys, dogs, mini-pigs, and rodents, including transgenic and chimeric animals modified to mimic specific disease attributes–must be deselected prior to determining that data required from a specific experiment can only be obtained through the use of a chimpanzee. Non-chimpanzee models in most cases sufficiently mimic the aspect of the disease (e.g., susceptibility, sustainability, progression) or disease pathways or targets, to the extent that they will provide sufficient data for the question being asked. The model system chosen does not need to replicate the complete pathophysiology of the disease/disorder being studied.

Species Differences in Absorption, Distribution, Metabolism, and Excretion (ADME)
Other species–such as monkeys, dogs, mini-pigs, and rodents, including transgenic and chimeric animals modified to mimic specific disease attributes–must be deselected by determining that ADME profiles do not adequately match the profile generated by humans.

Other species–such as monkeys, dogs, mini-pigs, and rodents, including transgenic animals modified to mimic specific disease attributes–must be deselected prior to determining that pharmacokinetic data (bioavailability, distribution, or metabolic data) obtainable from these species are significantly less suitable than data that are expected to be obtained from chimpanzees. For example, if a species fails to convert a pro-drug (inactive drug) to the active moiety, that species would be unsuitable as a toxicology species.

The standard in vitro (e.g., microsomal) model must be deselected when metabolism and pharmacokinetic data must show qualitative or substantial quantitative differences, and incremental differences are not considered sufficient.

Species Differences in Vehicle Tolerability
Other species–such as monkeys, dogs, mini-pigs, and rodents, including transgenic and chimeric animals modified to mimic specific disease attributes–must be deselected by determining that the test article is unable to be formulated in a vehicle tolerated by these models. In these limited cases, the chimpanzee may be justified if the formulation is tolerated in the chimpanzee and if testing in humans is not ethically possible (see below).

Species Differences in Response to Test Article Tolerability

Deselecting other species—such as monkeys, dogs, mini-pigs, and rodents, including transgenic and chimeric animals modified to mimic specific disease attributes—must be data driven. These data can be derived from in vitro studies (e.g., test articles demonstrated to be potent COX2 inhibitors or SSRIs are contraindicated in dogs, and some antimicrobials are contraindicated in rabbits and guinea pigs) if there is strong historical evidence of compound class intolerability.

Poor tolerability is justification for not using other species only if it precludes assessment of other relevant toxicities (e.g., if emesis precludes achieving adequate systemic exposure). If the basis for intolerability of a test article is clinically relevant, it may be a reason for selection rather than deselection of a non-rodent or non-human primate (NHP) species.

Pharmacology

The standard non-rodent and NHP species must be deselected if there is a lack of pharmacologic response (demonstrated inactivity) in these animals. In these cases, the chimpanzee may be justified if there is scientific evidence that pharmacological activity will occur in chimpanzees and those specific safety concerns of exaggerated pharmacology need to be characterized in animal toxicity studies. If other species—such as monkeys, dogs, mini-pigs, and rodents, including transgenic animals modified to mimic specific disease attributes—have pharmacological sensitivity that precludes testing at adequate multiples of clinical exposure, use of chimpanzees may be justified if toxicity studies in chimpanzees could achieve significantly greater exposure. However, if safety concerns of exaggerated pharmacology can be adequately characterized in other species—such as monkeys, dogs, mini-pigs, and rodents, including transgenic animals modified to mimic specific disease attributes—pharmacological responsiveness of chimpanzees is not necessarily a factor in species selection.

Immunogenicity

Other species—such as monkeys, dogs, mini-pigs, and rodents, including transgenic animals modified to mimic specific disease attributes—must be deselected if there is a scientifically based expectation for significant antigenicity for test articles not intended to be immunogenic. Vaccine research and development requires an appropriate immunogenic response to the vaccine and/or to an adjuvant, which in some cases may necessitate the use of chimpanzees, if human experiments cannot be ethically performed (see below).

Availability of Test Article or Cost of Species

A limited supply of the most suitable experimental animal or individual cost of the proposed species is not a justification for deselecting the standard non-rodent or NHP species.

Assessing Whether the Research Can Be Performed on Human Subjects

As the criteria regarding necessity outline, chimpanzee research is not necessary if it can be ethically performed on humans. Standard arguments about protection of human subjects require that there be an acceptable balance of the risks and potential benefits of proposed research, that the distribution of the risks and benefits are equitable (higher risk research can be justified when the potential therapeutic benefits accrue to the subjects themselves), and that the subjects are voluntary and informed of potential liabilities during their decision making. Relevant examples of critical human health-related research that would not meet human subjects' protection standards include trials that intentionally expose subjects to untreatable infectious diseases and exposure trials to hazardous substances that pose significant health risks without prospect of benefit.

When research on humans is justified, federal policies on protection of human subjects impose limits, including for research on subjects who cannot consent for themselves. Subparts of the federal regulations concerning research on human subjects also impose clear limits on acceptable research on children and prisoners (HHS, 2005). These include restrictions on research that poses greater than minimal risk to subjects; such research cannot be approved unless it has the potential for offsetting therapeutic benefit to the subjects themselves.

These standards and additional protective restrictions mean that more research may take place using animal models than would otherwise be the case if additional risks to human subjects were deemed acceptable.

Assessing Advancements to Treat Potentially Life-Threatening or Debilitating Conditions

The standard non-rodent and NHP species may be deselected if it can be demonstrated that forgoing the use of chimpanzees for the research in question will significantly slow or prevent important advancements to treat potentially life-threatening conditions in humans or debilitating conditions that have a significant impact on a person's health, and thus slow or prevent important advancements for the public's health. This assessment is based on the potential impact on human health and potential to improve well-being, which can be partially assessed by the burden of the disease or disorder. The committee notes that for emerging infectious diseases and biodefense-related threats, this information may not exist for low-probability, high-consequence threats.

Criteria for Use of the Chimpanzee in Comparative Genomics and Behavioral Research

As previously discussed, research using the chimpanzee raises unique ethical issues because of its genetic proximity to human beings and highly developed cognitive and social skills. Therefore, based on the principles previously defined, the committee developed the following criteria to guide its assessment of NIH-funded comparative genomics and behavioral research using the chimpanzee:

1. Studies provide otherwise unattainable insight into comparative genomics, normal and abnormal behavior, mental health, emotion, or cognition; and
2. All experiments are performed on acquiescent animals, in a manner that minimizes pain and distress, and is minimally invasive.

Specific and full scientific justification for the continued and future use of the chimpanzee must meet the above criteria, as well as the housing/maintenance requirements described earlier in the document. This assessment should be applied prospectively on a study-by-study basis.

Assessing the Objectives of the Project

The review of research projects on a study-by-study basis must demonstrate that the primary objective of the research is to provide otherwise unattainable, specific insight into human evolution, normal and abnormal behavior, mental health, emotion, or cognition. Research may be either basic or applied, but must be consistent with the mission of the NIH "to seek fundamental knowledge about the nature and behavior of living systems and the application of that knowledge to enhance health, lengthen life, and reduce the burdens of illness and disability" (NIH, 2011).

The committee recognizes that most behavioral research differs fundamentally from biomedical research in the sense that mental or behavioral disorders (with few exceptions) cannot be modeled explicitly using chimpanzees. This is because the naturally occurring prevalence of such disease is likely to be low if compared to what is observed in human populations, thus precluding reasonably sized studies using chimpanzees. Some conditions (e.g., depression or post-traumatic stress syndrome) may be inducible in chimpanzees, but likely only using procedures that would be judged unacceptably invasive. This is especially true inasmuch as other animals, including other nonhuman primates, have been used to model these disorders. It is for the forgoing reasons that the majority of comparative genomics or behavioral studies using chimpanzees have focused on continua of behavioral and developmental phenomena from normal to abnormal, taking advantage of similarities in behavioral and brain complexity that mark chimpanzees and humans apart from virtually all other species.

Assessing Animal Acquiescence and Distress

Comparative genomics and behavioral research should only be performed on acquiescent animals and in a manner that minimizes distress to the animal. Evidence of acquiescence includes situations in which animals do not refuse or resist research-related interventions and that do not require physical or psychological threats for participation. In addition, only minimally invasive protocols should be performed. Examples of minimally invasive procedures include behavioral observation and the introduction of novel objects to the living area. In performing some comparative genomics or behavioral research, it also may be necessary to temporarily isolate an animal from its social group to perform behavioral tasks or for anesthesia. It is anticipated that anesthesia may be necessary for noninvasive imaging studies, the collection of biological samples (including blood, skin, adipose, or muscle) that do not involve surgical invasion of body cavities, the implantation of radio transmitters to measure autonomic nervous system function or physical activity, and the use of biosensors for recording central nervous system responses in freely moving animals. Whenever possible, anesthesia for comparative genomics or behavioral purposes should coincide with scheduled veterinary examination. Research on elderly or infirm animals in particular should take full advantage of anesthesia performed as part of routine veterinary care. It is recognized, however, that some study protocols may require that animals be anesthetized apart from veterinary examinations. The annual occurrence of such episodes of anesthesia should be minimized in number and the length of time the animals are sedated, consistent with accepted veterinary practice, including post-procedure analgesia as required. In all instances, anesthesia protocols should be designed to ensure that effects on the central nervous system or other organs are transient, and anesthesia for research purposes only should be avoided when possible in elderly or infirm animals. When animal protocols for anesthesia are not available, protocols used for human patients under similar circumstances may guide the choice of procedures.

Finally, when temporary removal from the social group is required for behavioral manipulation or anesthesia, animals must be handled in a manner that minimizes stress. Successful strategies have included positive reinforcement training that allows animals to be called by name or otherwise enticed to leave their habitual setting to engage in research procedures.

REVIEWING THE NECESSITY OF CURRENT CHIMPANZEE RESEARCH

The following case studies are meant to demonstrate how the committee envisions its criteria for the use of chimpanzees in research might be employed. In each case, the committee reviews the current use of the chimpanzee against the criteria and makes a determination of whether or not the research should be continued or prohibited. It is important to note that the committee is not reviewing any specific research grant, but rather the larger body of research in each area. As reviewed previously in the report, chimpanzees are used in multiple research areas (see Figure 1). Based on the propensity of current research, the committee chose to assess the necessity of the chimpanzee in areas of research where there is significant on-going research or a potential for significant research. The committee assessed the following research areas: monoclonal antibodies, RSV, hepatitis C virus (HCV) antiviral drug development, HCV vaccine development, comparative genomics, cognition, and neurobehavioral function. Other areas, for example, malaria research, have limited on-going studies using the chimpanzee. From 2001-2010 there were only two studies that were done in the field of malaria, both currently still funded. For this reason, the committee chose not to use this and similar areas for case studies. However, the use of the chimpanzee in this and other research areas not reviewed by the committee can be assessed by using the same criteria.

Monoclonal Antibodies

Background

Currently, two separate uses of monoclonal antibodies rely on the chimpanzee. These are the production of chimpanzee monoclonal antibodies and preclinical safety testing of monoclonal antibodies prior to their introduction into humans. The development of monoclonal antibodies for use in any laboratory or clinical application follows the groundbreaking methods pioneered by Georges Köhler and César Milstein in the mid-1970s (Köhler and Milstein, 1975). Köhler and Milstein developed robust cell culture methods to immortalize individual B cells and thus create clonal cell lines that produce one type of antibody, hence the term "monoclonal antibody." The ability to produce essentially unlimited supplies of a unique monoclonal antibody provides a powerful technological platform for the generation and use of a wide range of affinity reagents in a myriad of applications.

In recent years the utility of having antibodies that bind to a single site on a molecule of interest has been expanded by the ability to produce affinity reagents using any of a series of in vitro molecular cloning methods (reviewed extensively over the years, but see de Marco, 2011; Demarest and Glaser, 2008; and Kneteman and Mercer, 2005, for recent comprehensive reviews). These approaches range from simple cloning of cDNA copies of the antibody mRNAs from immortal B cells, which allows the production of the monoclonal antibody in other cells and in vitro systems, to complete synthetic methods that identify individual binding domains from pools of expression vectors. The sequences that encode the binding domains can be expressed to produce a wide range of affinity reagents. It is now common to place the antigen interaction domains in antibody sequences from any organism, including humans, or in any antibody subtype, allowing the functional activities to be selected to achieve the best results. The antigen binding sites can be fused to other domains to make chimeric molecules that allow the production of reagents that bind to an antigen of choice and bring essentially any functional activity to the location of the antigen. These methods allow researchers to tailor affinity reagents to fulfill a wide range of desired activities. While monoclonal antibodies are still most commonly made by immunization of animals and immortalization of their B cells, synthetic or semi-synthetic methods are gaining increasing application.

Development of Chimpanzee Monoclonal Antibodies

For slightly over a decade researchers have been using the chimpanzee for the production of monoclonal antibodies (Altaweel et al., 2011; Chen et al., 2006a, 2006b, 2007b, 2009; 2011b, 2011c; Goncalvez et al., 2004a, 2004b, 2007, 2008; Men et al., 2004; Schofield et al., 2000, 2002, 2003). Typically these monoclonal antibodies are prepared by cloning antibody-encoding cDNAs from immunized chimpanzees. In brief, one or a small number of chimpanzees are injected with an immunogen of interest. Immunogens that have been used for successful monoclonal antibody production have included such agents as inactivated human viruses or bacterial toxins. At a chosen interval after the final boost, a bone marrow sample is collected from the chimpanzee. Lymphocytes are purified from the bone marrow samples, RNA is isolated, and cDNA is prepared for cloning in various expression vectors. Coding sequences that express protein fragments can bind to the desired immunogen and are then isolated. In most procedures the chimpanzee-coding region for antigen-binding domains are cloned as chimpanzee/human chimeric antibodies and used for subsequent experiments.

It has been suggested that this approach provides two potential advantages over monoclonal antibody production in other species. First, because the antibody protein sequences between the chimpanzee and the human are so similar (Ehrlich et al., 1990), further subcloning and humanization of the chimpanzee antibody sequences are not needed, and the resulting antibodies can be used directly in humans without further work. Second, because the immune responses of the chimpanzee and the human are so similar, it is likely that chimpanzees would mount immune responses that are similar to analogous immune challenges seen in humans. The chimpanzee/human chimeric monoclonal antibodies produced in these manners have proven to be effective in both in vitro and in vivo assays to neutralize infectious viruses or to block the action of bacterial toxins.

Criteria 1: Alternative Models

It is possible to develop monoclonal antibodies with these types of binding specificities in species other than chimpanzees. As is commonly done, these binding domains can readily be converted into fully humanized antibodies (see Nelson et al., 2010, and the references within for a review of this procedure and its common use in antibody therapeutics).

Monoclonal antibodies prepared in other species with properties similar to the chimpanzee antibodies are already described in the literature (reviewed by Marasco and Sui, 2007). Further, genetic humanization of the immunoglobulin locus in mice allows for rapid and high throughput production of fully human antibodies. For example, Regeneron Pharmaceuticals has created the so-called VelocImmune mouse by directly replacing mouse antibody gene segments with their human counterparts at the same location (Valenzuela et al., 2003). Alternatively, human antibodies can be induced in human xenotransplantation models (Becker et al., 2010). While the chimpanzee is clearly capable of making an effective humoral response to these immunogens, there seems to be no unique properties to the resultant antibodies to suggest that the continued use of the chimpanzee is required.

Finding

The committee finds that the continued use of chimpanzees for the production of monoclonal antibodies does not meet the suggested criteria for the use of the chimpanzee in biomedical research. Production of monoclonal antibodies following immunization in other species or through in vitro synthetic methods is equally powerful for the generation of such reagents. There appear to be no obvious reasons to suggest that the immunogenic regions of the antigens used for monoclonal antibody production in the chimpanzee are unique to this species. Neutralizing antibodies appear in other species in high frequency, and therefore it seems likely that antigen-binding domains seen in species other than the chimpanzee can be identified and used for the production of these reagents. The humanization of these antibodies should be similar in scope and difficulty to the approaches used with the chimpanzee, and the resulting reagents should be equally useful in humans. No added time savings are inherent in approaches compared to work in other species.

Safety Testing of Monoclonal Antibody Therapies

Monoclonal antibodies used in treatment of human disease bind to a carefully chosen antigen, often a protein, and through this interaction interfere with a cellular process that underlies disease development. Therapeutic monoclonal antibodies have become important front-line treatments for a wide range of human diseases and clinical procedures, including inflammation, autoimmunity, cardiovascular disease, cancer, macular degeneration, and transplantation. The first monoclonal antibodies approved for clinical use were introduced in the mid-1980s. The pace of FDA approval of monoclonal antibody-based therapies continues to increase—one treatment was approved in the 1980s (FDA, 1986), 7 in the 1990s, and 18 in the 2000s (An, 2010; Beck et al., 2010; Nelson et al., 2010; Reichert et al., 2005; Reynolds, 2011). Given the number of current clinical trials that are exploring new uses of monoclonal antibodies, it is likely that the introduction of novel therapies that rely on monoclonal antibodies will continue at least at this level in the future, and it is reasonable to speculate that the rate of FDA approval for new therapies may increase significantly over time.

Criteria 1: Alternative Models

When developing monoclonal antibody therapies for human clinical use, it is important to determine what, if any, unexpected effects these treatments might provoke in humans (see ICH Harmonized Tripartite Guideline, 2011, for regulatory practices and Chapman et al., 2009, and Tabrizi et al., 2009, for discussions of the process and needs for preclinical safety testing). Good preclinical models should mimic the biological effects of introducing the monoclonal antibody into humans and thus would provide predictions of any unexpected effects in humans. Issues that are important for the measurement of safety include the display of target molecules with analogous binding sites for the monoclonal antibody therapeutics, immune responses that are as similar to the human as possible, similar kinetics of monoclonal antibody presentation and clearance, and minimal immune response to the monoclonal antibody. The chimpanzee provides this close relationship, and has often been used as a model (Chapman et al., 2007).

Preclinical tests in the chimpanzee may lead to adverse events, and these adverse events may arise from three sources. First, the antibody could bind with the intended targeted protein, but the target may have unknown roles in the body that are unrelated to its disease-causing effects, thus giving on-target toxicity. Second, the monoclonal antibody may bind to proteins other than the intended target, and these interactions could give rise to unwanted side effects, yielding off-target toxicity. Third, analogous types of on- or off-target toxicities could arise from functional domains on the monoclonal antibody other than the antigen-binding domain. Other models, such as other monkey species, have not proven to be as effective for detecting such toxicities, and over time it has become common to test for such unwanted effects using appropriately monitored and carefully sized trials in the chimpanzee. Undesired results in chimpanzee safety tests have led to the termination of a number of monoclonal antibody programs before they have advanced to clinical tests in humans, presumably saving unwarranted human suffering in the process (Abee, 2011a; Reynolds, 2011). In addition, there are also rare examples of monoclonal antibodies that have been tested directly in humans without previous chimpanzee safety tests and that have caused severe and undesired responses in humans (see Eastwood et al., 2010, for a potential biological explanation of one such undesired response). Therefore, the use of the chimpanzee for safety tests has proven to be valuable, and such studies have been used to protect human health.

Although safety trials for monoclonal antibody therapies continue to be performed in the chimpanzee, the committee also has noted that in recent years there is a trend in many groups to avoid its use. This trend is driven both by advances in monoclonal antibody technology and by changes in how potential monoclonal antibody treatments are first introduced into humans.

There are currently four methods in use that lessen the need for safety tests in the chimpanzee: (1) genetic engineering of the target protein in rodents; (2) selection of antibodies that recognize target epitopes shared across species; (3) selection of multiple antibodies that can serve as surrogates for responses; and (4) microdosing in humans (Chapman et al., 2007; Reynolds, 2011).

The first of these approaches relies on expressing the target protein in a rodent, expressing the target epitope's ortholog in the rodent, or developing mice with xenotransplanted human tissue. This is an approach that offers some benefits, but there is considerable worry that the target protein may not function identically in the rodent, and other functional domains on the monoclonal antibody may not be recognized in an identical fashion in the rodent compared to the human. Since much of the potential response to monoclonal antibody treatment cannot be mimicked by this method, it has a limited potential to change the necessity for chimpanzee use. Nonetheless, this is a useful experimental approach and can help guide researchers to potential problems. By itself, however, this approach does not significantly change the need for chimpanzee research.

Two other approaches to lower the need for chimpanzees in safety testing rely on changes in how monoclonal antibodies are chosen for potential clinical development. As mentioned above, the development of various recombinant antibody methodologies has dramatically expanded the range of properties that can be selected or developed during antibody creation. In one useful approach, researchers select monoclonal antibodies that bind to the target antigen at sites found both in the human and in other species beyond the chimpanzee, often in another NHP (Reynolds, 2011). In these cases, the safety of interfering with the activity of a disease-specific protein can be tested in species other than the chimpanzee. If this species has other features that mimic the human, confidence in the safety profile of a preclinical candidate being considered for human studies is raised. There is still some considerable concern about how well the preclinical model mimics the human, and it is commonly argued that even good results in such tests cannot ensure how human tests will proceed. Here, as in other cases, safety studies in more than one species raise confidence about prediction of response in the human.

In a third approach, researchers choose two or more monoclonal antibodies that bind to the same target protein. These antibodies are frequently called surrogate antibodies. With two or more antibodies that bind to the same target antigen, on-target effects can be established by comparing the responses in dose-escalating safety studies. These studies are performed in preclinical safety models. On-target effects can be identified as those that are common to all antibodies, while undesired effects are specific to one of the agents. While surrogate antibodies may not have all of the best properties of a true clinical development candidate, studying the responses to multiple agents can increase the confidence of the potential safety profiles of monoclonal antibodies prior to introduction into humans.

Criteria 2: Testing on Human Subjects

A fourth approach that may lower the dependence on safety testing in the chimpanzee relies on microdosing in humans. Monoclonal antibody treatments, which have previously shown good Pharmacokinetics/Pharmacodynamics (PK/PD) and toxicology results in preclinical studies in other models, can be tested for safety directly in humans using microdosing schedules, such as using minimal anticipated biological effect level strategies (see Muller et al., 2009, for a careful review of the use of microdosing strategies). Starting with very low doses enables clinical researchers to carefully monitor for any unexpected side effects in settings where adverse events can be detected before serious harm is done to the patient. These microdosing approaches can be teamed with the introduction of radiolabeled tracer preparations of the monoclonal antibody to follow the in vivo localization of the antibodies and thus potentially link side effects to particular organ sites for further studies.

Finding

These approaches—use of genetically engineered rodents, directed strategies to select monoclonal antibodies with broader binding specificities across species, use of surrogate antibodies, and different methods to introduce antibodies into humans—combined with the recognition that the FDA does not require safety testing of new monoclonal antibody therapies in chimpanzees, promise to provide a series of methods that can be used to protect human safety while avoiding use of the chimpanzee.[9] Therefore, the committee finds that use of these methods, often in combination, can make the chimpanzee largely unnecessary in the development of future monoclonal antibodies therapies.

Not all companies and few academic laboratories have fully adopted monoclonal antibody approaches, such as recombinant antibody production, that allow the selection of monoclonal antibody therapeutic agents that meet these more defined criteria. Therefore, there may be a limited number of monoclonal antibodies currently in the development pipeline that may require the continued use of chimpanzees. For these specific cases, the use of the chimpanzee should be assessed against the committee's criteria for biomedical research. In addition, the NIH should be expeditious in supporting the development of broadly accessible recombinant technologies for development of novel therapeutic monoclonal antibodies.

Respiratory Syncytial Virus

Prevalence and Treatment Options

Respiratory syncytial virus is a pneumovirus initially described as chimpanzee coryza agent in 1955 and renamed "respiratory syncytial virus" in 1956 following virus isolation from infants with bronchiolitis

---

[9]It is important to note that if data from safety testing in the chimpanzee are presented as part of an investigational new drug application, the FDA requires that these studies reach appropriately high standards to contribute to the prediction of eventual safety in humans.

and pneumonia (Blanco et al., 2010; Chin et al., 1969; Wright and Piedimonte, 2011). Overall, RSV is the most common cause of acute lower respiratory tract infections and bronchiolitis in children under the age of 5 years (Krilov, 2011; Nair et al., 2011). Nearly all children have been infected with RSV at least once by age 2, with a large percentage of infants infected during their first RSV season (Fulginiti et al., 1969; Weisman, 2009; Wright and Piedimonte, 2011). RSV is believed to account for 85 percent of bronchiolitis and 20 percent of pneumonia cases globally, with 1 in 200 infants requiring hospitalization (Nair et al., 2011). Today, RSV is the leading cause of hospitalizations in U.S. children less than 1 year old, with an estimated 100,000 to 126,000 infants hospitalized each year due to bronchiolitis (Krilov, 2011; Wright and Piedimonte, 2011). In addition to children, RSV is reported to have similar rates of hospitalization and mortality in the elderly (Shadman and Wald, 2011). Another at-risk population includes patients who have undergone hematopoietic stem cell transplantation, with RSV affecting approximately 2-17 percent of these transplant recipients (Shah and Chemaly, 2011).

Current treatment and prevention options for RSV are very limited. Treatment for hospitalized infants and children primarily includes supportive care, but may also include administration of α-adrenergics or corticosteroids (Krilov, 2011). For severe RSV-induced lower respiratory tract infections or where there is a high risk of mortality, Ribavirin is the only licensed antiviral currently on the market. Its use is limited due to factors such as minimal clinical benefits and high cost (Krilov, 2011; Wright and Piedimonte, 2011). Palivizumab, the only approved prophylactic RSV drug, is a humanized monoclonal antibody that is administered intramuscularly every 30 days during RSV season. It is only used to reduce severity and morbidity in high-risk populations, including premature infants less than 35 weeks of gestation, infants with chronic lung or congenital heart disease, or infants and children with lung abnormalities such as cystic fibrosis (Nair et al., 2011; Shadman and Wald, 2011). Palivizumab is not commonly used in low-risk populations due to the high cost of treatments and limited evidence of clear benefit in these populations (Krilov, 2011; Prescott et al., 2010; Shadman and Wald, 2011).

Criteria 1: Alternative Models

Multiple cell lines from human and animal sources are currently used in preclinical research of RSV, including differentiated normal human bronchial epithelial (NHBE) cells (DeVincenzo et al., 2010; Tayyari et al., 2011). Recently bronchial epithelial cells derived from human lung adenocarcinoma were found to be susceptible to RSV infection and release infectious virus similar to NHBE cells, indicating a new cell line for potential use (Harcourt et al., 2011). The use of cell culture has limitations, including the inability to replicate tissue organization and disease manifestations, such as respiratory infection in the case of RSV. Therefore, animal models of human RSV (hRSV) provide an important link between mechanistic cell culture research and human clinical trials.

No single animal model, including the chimpanzee, reproduces all aspects of RSV infection. Chimpanzees are susceptible to infection with hRSV with replication of the virus in high titers in the upper and lower respiratory tracts (Murphy et al., 1992). The chimpanzee also has the same body temperature as humans, unlike other animal models, which is important when investigating the degree of attenuation of candidate temperature-sensitive vaccine strains (Murphy et al., 1992). Seronegative chimpanzees can serve as surrogates for seronegative infants, the target population for vaccines. However, identification of animals that are seronegative can be difficult since the infection commonly spreads between animal handlers and chimpanzees (Murphy et al., 1992). The inability of chimpanzees to develop lower respiratory diseases such as bronchiolitis and pneumonia, the limited availability of specific reagents, and large size are noted disadvantages of this model (Bem et al., 2011; Graham, 2011). Alternative models for RSV research, including sheep, cotton rats, and mice, have significantly reduced the use of chimpanzees in RSV basic research; the last published research paper to use the chimpanzee appeared in 2000.

Sheep are susceptible to hRSV infection; neonatal lambs develop upper respiratory tract disease, bronchiolitis, and mild pneumonia. In addition to showing clinical signs of the disease, the respiratory tracts of sheep share many structural and developmental features with humans, unlike rodents (Bem et al., 2011). The limited availability of immunological reagents and other molecular tools, along with animal size, are two disadvantages of this model. Cattle provide another useful model of RSV infection because bovine RSV (bRSV) shares many common characteristics with hRSV, including respiratory tract infections, increased susceptibility of the young, and seasonal outbreaks (Byrd and Prince, 1997). Like sheep, however, the limited availability of reagents and animal size are disadvantages of this model. Currently the most commonly used animal models for hRSV are mice and cotton rats. The cotton rat is a semipermissive model of hRSV replication that develops mild to moderate bronchiolitis or pneumonia following exposure. Studies have demonstrated that this model can develop vaccine-enhanced disease at all ages and that the disease is similar to that seen in humans (Mohapatra and Boyapalle, 2008). However, cotton rats do not show clinical signs of respiratory tract disease (Bem et al., 2011; Graham, 2011). Mice, specifically the BALB/c inbred strain, develop lower respiratory tract disease symptoms following infection along with signs of illness such as weight loss (Bem et al., 2011; Domachowske et al., 2004; Nair et al., 2011). In addition, the immunohistopathology of RSV infection in mice resembles that of human infection (Mohapatra and Boyapalle, 2008). The ability to develop transgenic mouse lines offers a distinct advantage to the mouse model in comparison to other models. However, the difference in innate and adaptive immune responses between mice and humans and the inability for hRSV to robustly replicate in mouse lung tissue and spread between the upper and lower airway are disadvantages of this model (Bem et al., 2011; Mohapatra and Boyapalle, 2008).

Together these alternative models to the chimpanzee for RSV research demonstrate both susceptibility to the human form of the virus and the ability to develop clinical signs of the virus, including bronchiolitis and pneumonia; therefore, they cannot be deselected from use. Their availability and suitability, along with increasing number of cell culture systems, indicates that the first criteria for the use of chimpanzees are not met in the case of RSV research.

Criteria 2: Testing on Human Subjects

RSV antiviral drug and prophylactic vaccine clinical trials progress from Phase I studies in adults to trials with seropositive children and then seronegative infants (Nair et al., 2011). While time consuming, clinical testing is possible on human subjects, but the development of novel vaccines may be limited by an inability to predict adverse reactions to vaccines, something that can be accomplished in chimpanzees. This obstacle may be overcome with the recent development of a human experimental infection model (DeVincenzo et al., 2010). Healthy adult volunteers were infected with RSV, causing a self-limited upper respiratory illness. The researchers then tested the safety and efficacy of a small interfering RNA (siRNA) antiviral, ALN-RSV01. This proof-of-concept trial suggests a mechanism for development of a challenge model for testing vaccines against RSV in the future. While this human infection model is new and has only been examined in one study so far, it suggests future avenues for immunogenicity, efficacy, and safety testing in human subjects. The development of a human challenge model suggests that RSV research can be performed ethically on human subjects; therefore, the use of chimpanzees does not meet the second criteria for biomedical research.

Criteria 3: Impact of Forgoing Chimpanzee Use

Forgoing the use of the chimpanzee will not significantly slow or prevent advancement of either therapeutic or prophylactic drugs for RSV and therefore, chimpanzee use does not meet the third criteria. This finding is based, in part, on both the availability of multiple non-human animal models that recapitulate several aspects of RSV disease and the ability to conduct proof-of-concept trials in a human model of infection. Currently three vaccines and two antiviral compounds are in clinical trials. MicroDose Therapeutx is in a Phase I trial of MDT-637, an inhalable small molecule antiviral (MicroDose Therapeutx, 2011). Alnylam is conducting a Phase II efficacy and safety trial of the siRNA antiviral ALN-RSV01 in RSV-infected lung transplant patients (Alnylam Pharmaceuticals, 2011). Novavax and MedImmune are in Phase I and I/IIa clinical trials of potential vaccines, respectively. Novavax is testing four different recombinant RSV-F formulations of NVX 757 01 in healthy adults (Novavax, 2011). MedImmune is testing two attenuated intranasal vaccines (MEDI-534 and MEDI-559) in children and infants (MedImmune LLC, 2011a, 2011b). In addition to these, at least seven other pharmaceutical companies have preclinical RSV programs in development for either therapeutics or prophylactics, including a trivalent anti-RSV-F nanobody (Nb ALX-0171) derived from immunized camels (Pharmaceutical Business Review, 2011).

Finding

The committee finds that currently, chimpanzee use for RSV research is not necessary. This finding is based on the inability of the use of the chimpanzee to meet the three criteria for biomedical research, the current state of research, availability of alternative models, and the large number of drug development efforts. However, the committee cannot completely eliminate the potential for a future need of this animal model. The development of a safe and effective vaccine would confer the greatest benefit on the most vulnerable of populations, infants under 6 months of age. The committee acknowledges that there are still barriers in the development of a prophylactic vaccine for RSV, including the need to immunize young infants who potentially may not respond to vaccines or have adverse reactions, possible interactions with other pediatric vaccines, or enhanced reactivity (Blanco et al., 2010). The chimpanzee may be required in the future for testing of novel vaccines because of the ability of the chimpanzee to serve as an early surrogate model for seronegative infants (Mohapatra and Boyapalle, 2008; Pollack and Groothuis, 2002; Weisman, 2009).

HCV Antiviral Drugs

Prevalence and Treatment Options

Hepatitis C virus currently infects 130-170 million people worldwide (WHO, 2011). In the United States, 3.2 million people are chronically infected with hepatitis C virus (Williams et al., 2011). More than 350,000 individuals die each year due to HCV-induced cirrhosis, end-stage liver disease, or hepatocellular carcinoma (Klevens and Tohme, 2010). Current therapy for patients chronically infected with HCV includes pegylated interferon α and ribovarin plus the recent addition of an HCV protease inhibitor, telaprivir (Incivek) or boceprivir (Victrelis). This regimen leads to viral cures in a high percentage of HCV-infected subjects, including those with more-difficult-to-treat genotype 1 infections common in the United States and Europe (Ilyas and Vierling, 2011). The two new NS3/NS4A protease inhibitors, telaprevir and boceprevir, were approved by the FDA in 2011 based on their effectiveness in in vitro culture systems (Sheehy et al., 2007) and then in large controlled clinical trials (Jacobson et al., 2011; Jensen, 2011; Kwo et al., 2010). Additional inhibitors of HCV's NS5A replication complex assembly factor and the NS5B RNA polymerase are currently in advanced clinical development, offering future hope for a highly effective, completely oral, and interferon-free therapeutic regimen for patients chronically infected with HCV (Ilyas and Vierling, 2011). The current HCV pipeline now includes four drugs in Phase III and 22 in Phase II of development.

Criteria 1: Alternative Models

Because of their unique susceptibility to HCV infection and the initial lack of in vitro culture systems, chimpanzees were particularly valuable during the early phases of HCV research. For example, molecular clones of HCV were first isolated from a cDNA expression library prepared with mRNA from a chimpanzee infected with non-A, non-B hepatitis virus (Choo et al., 1989). More recently, the use of chimpanzees has declined as both HCV replicons (Lohmann et al., 1999) and fully infectious HCV molecular clones (Lindenbach et al., 2005; Wu et al., 2005; Zhong et al., 2005) were identified, enabling the establishment of in vitro culture systems. Various animal models (reviewed in Boonstra et al., 2009) have also emerged, including immunotolerized rats containing human hepatocytes (Wu et al., 2005); immunodeficient mice with heterotopic human liver grafts (Galun et al., 1995); SCID mice expressing the urokinase plasminogen activator transgene that destroys the endogenous mouse liver, permitting xenotransplantation of human hepatocytes (Mercer et al., 2001; Meuleman et al., 2005); and genetically humanized, immunocompetent mice containing human surface receptors required for HCV entry (Dorner et al., 2011). While culture systems are not yet available for genotypes 3, 4 and 6, the replicon system could be used to screen inhibitors against the protease and polymerase, but not NS5A. However, it is very likely that infectious molecular clones will soon emerge for these genotypes—a blueprint for their development now exists—and further alternative small animal models supporting the growth of these viruses will also likely progress rapidly.

The currently available experimental systems, coupled with the challenges inherent to chimpanzee experiments, including limited numbers of animals and high costs, have resulted in a steady deemphasis of the chimpanzee model in HCV antiviral drug design and development. For example, both boceprevir and telaprevir were developed and approved without the use of chimpanzees; instead, preclinical experiments were conducted in mice, rats, rabbits, dogs, and cynomolgus monkeys (EMEA, 2011b; Vertex Pharmaceuticals Incorporated, 2011). However, a few companies continue to use previously HCV-infected chimpanzees (Chen et al., 2007a; Olsen et al., 2011).

Criteria 2: Testing on Human Subjects

Chimpanzees have been used to establish PK/PD relationships of candidate drugs and to assess antiviral activity in vivo. An ethical alternative to performing PK studies in chimpanzees is now available. Specifically, Phase 0 studies can be performed in consenting humans involving microdosing of a drug candidate (Ings, 2009). Such microdosing studies involve the administration of the drug at very low, subtherapeutic amounts that are unlikely to produce toxic side effects, followed by monitoring of drug distribution and clearance using highly sensitive bioanalytical methods. Drugs with unacceptable pharmacokinetic profiles can be rapidly excluded from further development. For Phase I toxicity studies or more advanced efficacy studies, consenting individuals chronically infected with HCV could be recruited. The use of humans for evaluation of these HCV antivirals is further supported by the fact that HCV infection in chimpanzees only partially recapitulates the clinical and laboratory features of HCV infection in humans. Specifically, hepatic disease is milder in chimpanzees (Bukh et al., 2001) and a greater fraction of these animals spontaneously clear the virus (Bassett et al., 1998). Chronic HCV infection in chimpanzees also does not generally result in hepatic fibrosis or cirrhosis (Bukh et al., 2001), and chimpanzees, unlike humans, fail to transmit HCV vertically from mother to infant (Zanetti et al., 1995). Finally, chimpanzees mount weaker neutralizing antibody responses to HCV than humans (Su et al., 2002; Thimme et al., 2002). The current pace of HCV drug development is testimony to the adequacy of human subjects for most of this work.

Criteria 3: Impact of Forgoing Chimpanzee Use

Forgoing the use of chimpanzees will not significantly slow the development of new HCV antivirals. Many new classes of HCV antivirals are already approved or in advanced clinical trials (Ilyas and Vierling, 2011). Progress in their development has been driven not by the availability of a chimpanzee model, but rather by the emergence of powerful in vitro culture systems supporting production and spread of fully infectious HCV virions and by the large number of HCV-infected patients who are willing to participate in clinical trials. Additionally, new small animal models are further reducing the need for chimpanzees in HCV antiviral drug development (Dorner et al., 2011; Galun et al., 1995; Mercer et al., 2001; Wu et al., 2005). As noted, the pharmaceutical industry is steadily moving away from the chimpanzee model for HCV drug discovery and development.

Finding

The committee finds that chimpanzees are not necessary for HCV antiviral drug discovery and development and does not foresee the future necessity of the chimpanzee model in this area.

Therapeutic HCV Vaccine

Prevalence and Treatment Options

Approximately one out of every 30-50 persons in the world is chronically infected with HCV (WHO, 2011) but, encouragingly, 15-30 percent of humans acutely infected with HCV succeed in clearing the virus (Wang et al., 2007). Many individuals exhibiting chronic HCV viremia lack specific T cell responses to the virus, including production of interferon-γ (Cooper et al., 1999). This finding suggests it may be possible to create a therapeutic HCV vaccine eliciting the type of missing immune response required for a sustained viral response and viral clearance (Houghton and Abrignani, 2005). A therapeutic vaccine would be given to individuals already infected with HCV in contrast to a prophylactic or preventive HCV vaccine that would be given to uninfected individuals who are at risk for infection. By redirecting the immune response and clearing the virus, a therapeutic vaccine could halt and potentially reverse progression of hepatic disease. A therapeutic vaccine also represents an attractive and cost-effective alternative to antiviral drugs in the management of patients with chronic hepatitis C infection. It could be particularly useful in patients who either are unable to tolerate interferon-α or fail to respond to this cytokine. The intrinsic sequence variation of HCV within each of its 6 recognized genotypes and more than 50 subtypes poses a major challenge to the successful development of a therapeutic HCV vaccine (Kurosaki et al., 1993).

Criteria 1: Alternative Models

Because of its tropism and growth requirements, HCV infection in vivo is limited to chimpanzees and humans. Small animal models involving implantation of human liver into immunodeficient mice (Galun et al., 1995) or engineering wild-type mice to express HCV entry receptors on hepatocytes (Dorner et al., 2011) have been developed. However, these mouse models are currently not appropriate and will require additional improvements for testing a therapeutic HCV vaccine where high-titer HCV infection in an immunocompetent host is required. Currently, chimpanzees and humans represent the only acceptable options for testing of an HCV therapeutic vaccine.

Criteria 2: Testing on Human Subjects

Subjects chronically infected with HCV are frequently used to test therapeutic HCV vaccine candidates. Therapeutic vaccine candidates are now being tested in humans without prior testing in the chimpanzee model (Halliday et al., 2011). The fact that chimpanzees produce weaker neutralizing antibody responses than humans (Thimme et al., 2002) and fail to respond to interferon like many humans (Lanford et al., 2007) argues that humans might represent a better system for testing therapeutic HCV vaccines.

Criteria 3: Impact of Forgoing Chimpanzee Use

The fact that therapeutic vaccine testing can be performed in consenting human subjects chronically infected with HCV without prior experimentation in chimpanzees indicates that forgoing the use of chimpanzees would have little or no impact. It is possible that direct testing in humans might accelerate development of an efficacious therapeutic vaccine for HCV.

Finding

The committee finds that chimpanzees are not necessary for development and testing of a therapeutic HCV vaccine.

Prophylactic HCV Vaccine

Prevalence and Treatment Options

HCV is an important cause of human disease—about 3.2 million Americans are chronically infected with hepatitis C virus, mostly from initial infections occurring years ago; however, there are about 17,000 new infections each year, according to the CDC (Williams et al., 2011). The U.S. incidence has fallen dramatically over recent years, but the disease remains a major problem worldwide, with 130-170 million infected. Persistent infection is common and can lead to liver fibrosis, cirrhosis, and hepatocellular carcinoma; hepatitis C has become the most common cause of liver failure and liver transplantation.

Rapid advances are being in made in the development of new therapeutics for subjects chronically infected with HCV, but an efficacious prophylactic vaccine against this virus has not yet been produced. Creation of such a vaccine will be especially challenging because of the great genetic and antigenic diversity manifested within HCV's multiple genotypes, subtypes, and quasi-species (Halliday et al., 2011).

Chimpanzees are highly susceptible to experimental HCV infection—in fact, the virus was initially identified by its transmission to chimpanzees, followed by molecular methods to detect viral RNA in infected chimpanzee plasma. The unique tropism of HCV for chimpanzee and human hepatocytes makes the chimpanzee model of experimental infection valuable for studies of pathogenesis, including mechanisms of persistence, and for development and testing of prophylactic HCV vaccine candidates by helping to identify those that are safe and efficacious.

The chimpanzee model could also provide important insights into the correlates of immune protection (Strickland et al., 2008); however, research is proceeding on prophylactic HCV vaccine development in the absence of testing in chimpanzees (Catanese et al., 2010; Garrone et al., 2011). Although the determinants of protective immunity against persistent infection with multiple strains have not yet been defined, studies of the outcome of vaccination and challenge of chimpanzees suggest that immune responses to envelop glycoproteins are important for protection, while responses to nonstructural proteins may be detrimental (Dahari et al., 2010; Houghton, 2011). Both neutralizing antibodies and T cell responses are likely to be important (Choo et al., 1994; Folgori et al., 2006; Meunier et al., 2011; Shoukry et al., 2003; Verstrepen et al., 2011). Of note, interpretation of many of these studies is complicated by the use of small numbers of animals (Bettauer, 2010), coupled with the fact that chimpanzees more effectively clear HCV infections than humans and are less likely to develop hepatic fibrosis, cirrhosis, or hepatocellular carcinoma than humans (Bassett et al., 1998, 1999; Boonstra et al., 2009; De Vos et al., 2002; Erickson et al., 2001; Thomson et al., 2003).

Criteria 1: Alternative Models

Chimpanzees and humans are the only two species that are susceptible to HCV infection. Currently, no other suitable animal models exist for evaluation of a prophylactic HCV vaccine. Although progress is being made in the development of various mouse models that can be infected with HCV, these do not allow evaluation of the human protective immune response against HCV. One model developed requires engraftment of human hepatocytes into the injured liver of an immunodeficient mouse, so HCV infection can be established, but the mouse is not capable of an adaptive immune response (Bissig et al., 2010). A second model involves ectopic implantation of human liver tissue into an immunocompetent mouse, so infection could theoretically occur, but any immune response would be murine in origin (Chen et al., 2011a). In the most recent model, transgenic mice have been engineered to express human HCV entry receptors, so infection can be established in an immunocompetent mouse and immune-mediated protection evaluated, but again, the immune response is murine (Dorner et al., 2011; Gewin, 2011). Likewise, no in vitro systems currently exist that display both HCV infectivity and the capability of an effective anti-HCV adaptive immune response. It is not known whether the recent identification of a canine hepacivirus, which is closely related to HCV and causes respiratory disease (Kapoor et al., 2011), will provide an additional relevant animal model system for vaccine testing and development.

Criteria 2: Testing on Human Subjects

Studies in consenting humans at high risk for HCV infection can be ethically performed to evaluate prophylactic HCV vaccine candidates, provided these vaccines are first shown to be safe and immunogenic in experimental animals such as mice and nonhuman primates. This type of human study will ultimately be required for any prophylactic HCV vaccine to gain licensure for widespread use. These studies require large numbers of subjects that are at increased risk of HCV infection. In developed countries, most new infections occur in people who inject drugs, a population that presents biological, methodological, and ethical challenges for vaccine trials (Maher et al., 2010).

One clear advantage offered by the chimpanzee model is the ability to infect the animals at a precise time following administration of a vaccine candidate. This could facilitate identification of promising vaccines and help define correlates of immunity and determine the durability of protection. However, a truly efficacious HCV vaccine must provide long-lasting protection within the human population, making such timing less important.

Criteria 3: Impact of Forgoing Chimpanzee Use

While ethical prophylactic vaccine studies in high-risk human populations can and ultimately must be performed, such trials are likely to prove challenging and time-consuming. Use of the chimpanzee HCV model of experimental infection could potentially speed identification of promising prophylactic HCV vaccine candidates for testing in humans, though the FDA does not have policies requiring data from chimpanzees for the development of any compound or vaccine (though will accept such data if submitted). However, differences in the pathogenesis of HCV infection in chimpanzees and humans with respect to immune responses, including weaker neutralizing antibody responses and higher rates of spontaneous viral clearance in chimpanzees, must be considered in judging the various vaccines. In addition, preclinical experiments using chimpanzees must be designed to include adequate numbers of animals for the generation of statistically meaningful results. Ongoing research that is proceeding without using chimpanzees may avoid these weaknesses, though such efforts are in their early stages.

Finding

The committee finds that while there are limitations to the current chimpanzee preclinical model, it has provided valuable knowledge for the development of prophylactic HCV vaccines. The committee is aware of progress on non-chimpanzee models that can be infected with HCV. Such models, if further improved, may reduce or obviate the need for the continued use of the chimpanzee for prophylactic HCV vaccine research. Moving directly to human trials in high-risk populations, without prior testing in chimpanzees, can be ethically performed and could lead to the development of an HCV prophylactic vaccine. After consideration of all of these facts, the committee was evenly split and unable to reach consensus on the necessity of the chimpanzee model, and on whether or how much the chimpanzee model would accelerate or improve prophylactic HCV vaccine development. Specifically, the committee could not reach agreement on whether a preclinical challenge study using the chimpanzee model was necessary and if or how much the chimpanzee model would accelerate or improve prophylactic HCV vaccine development.[10]

Comparative Genomics

Molecular genetics and comparative genomics hold enormous potential for developing biomedical therapies as well as for a more basic understanding of the origins of our own species. However, true genomic advances require two components beyond genetic material: (1) phenotypes

---

[10]As elaborated in the case study discussion, the committee could not agree on the necessity of the chimpanzee for research involving the development of a prophylactic HCV vaccine. In summary, the disagreement centered on whether chimpanzee testing is a necessary step in the path to human trials of candidate vaccines.

Some members of the committee held the view that chimpanzees provide the only available challenge model for testing a candidate vaccine and that without the use of chimpanzees (1) important data regarding the immunogenicity, protective efficacy, and safety of candidate vaccines would be foregone; (2) studies in populations of humans at high risk of HCV infection are likely to be difficult based on population demographics and currently available HCV treatment options; and (3) some candidate vaccines of limited promise might make their way into human trials, at the cost of additional time and resources.

An equal number of committee members held the view that (1) rodent and other rapidly-developing alternative models can provide sufficient immunogenicity and safety data to proceed to human efficacy trials without the need for prior studies in chimpanzees, that chimpanzee data is not always predictive of vaccine toxicity or efficacy in humans, and that use of the chimpanzee model is frequently complicated by the lack of a sufficient number of animals to generate statistically significant results; these committee members felt that foregoing chimpanzee models may actually spur greater attention to developing more tractable experimental alternatives; (2) studies in populations of humans at high risk of HCV infection studies can be designed and carried out successfully; and (3) the likelihood and length of any possible delay in vaccine development caused by foregoing chimpanzee research is difficult to assess, and human trials are required whether or not research proceeds using the chimpanzee during the course of vaccine development.

It is important to note that there was a consensus among the committee that human trials of candidate vaccines could be designed and performed ethically with or without data from chimpanzee research.

that can be linked to underlying genes, gene expression, or genetic processes; and (2) comparative studies across species to elucidate the origin and potential impact of genetic variation. The sequencing of the chimpanzee genome (Mikkelsen et al., 2005) in addition to the mouse (Waterston et al., 2002), the macaque (Gibbs et al., 2007), and other species, has shown that changes that set human beings apart from other species not only involve amino acid substitutions, but also to a large extent relate to minor alterations in regulatory regions, gene transcription, and gene expression (Marques-Bonet et al., 2009). However, the ability to apply the new wealth of genomic and genetic information to health and behavioral problems is impeded to the extent that developmental, physiological, behavioral, cognitive, and other phenotypic information is absent for comparator species, especially the closest and in some ways most informative taxon, the chimpanzee.

The development of advanced sequencing techniques provides access to another tool that investigators can use to further examine tissues that are obtained either from biopsy or necropsy. For example, transcriptional assessment can be applied to tissues in chimpanzees with different life histories and disease experience as well as to multiple tissues from the same chimpanzee. Analysis of the resulting gene expression profiles could greatly enhance our understanding of the biological pathways that are activated among individuals that have been subjected to specific life experiences and disease states. While such information is slowly becoming available from humans and other species, the systematic study of gene expression in the current NIH-supported population of chimpanzees may comprise an important source of biomedical and behavioral knowledge. Moreover, the mechanisms underlying environmentally induced alterations in gene expression are also becoming better understood and applied to chimpanzees. Specifically, epigenetic regulation of gene expression through DNA methylation or histone modification can now be readily evaluated in a wide variety of tissues and in blood. The general understanding of functional similarities and differences between chimpanzees and human proteins will be informed by better understanding of the factors affecting epigenetic regulation of genes and the resulting patterns of gene expression in chimpanzees, as compared to humans. Finally, insight through the examination of alternative splicing may also yield valuable information. For example, 6-8 percent of the proteins they examined exhibited profound differences in splicing between chimpanzees and humans and hypothesized that alternative splicing is an important source of the differences between humans and chimpanzees.

Forkhead Box P2 Gene Function in Chimpanzees and Humans

Chosen here for case evaluation in comparative genomics is an investigation that uses material from chimpanzees and humans to evaluate transcriptional regulation of the area of the central nervous system (CNS) that appears to encode language development—the preeminent human characteristic (Konopka et al., 2009). The human capacity to generate spoken language and the ability to understand and apply language to complex problems is a hallmark human feature, the origins of which remain relatively poorly known. Over the past several years, a number of investigators have produced various types of data demonstrating that the "forkhead box P2" or FOXP2 gene has important effects on the development of language skill (Enard et al., 2002). FOXP2 is a transcription factor, a gene that codes for a protein that binds to DNA and functions by turning on and turning off numerous other genes.

Furthermore, comparisons of the FOXP2 gene sequence in humans, chimpanzees, and other primates indicate that the human FOXP2 gene has undergone significant, rapid, and recent evolutionary change (Enard et al., 2002). This pattern has led researchers to infer that changes occurring in FOXP2 in human ancestors—after their divergence from the ancestors of modern chimpanzees—may help explain the evolution of the human capacity for language. Equally important from a biomedical perspective, mutations in FOXP2 have been associated with speech and language disorders (Lai et al., 2001; MacDermot et al., 2005).

Criteria 1: Studies Provide Otherwise Unattainable Insight

Using human neuronal cell lines that express either the human or chimpanzee forms of FOXP2, the investigators examined the function of the FOXP2 protein (Konopka et al., 2009). The results indicated that the human form of the protein had significantly different effects on the expression of many other brain-expressed genes, compared with the chimpanzee form of FOXP2. In addition, the investigators examined RNA isolated from both human and chimpanzee brains and showed that many of the genes that were differentially regulated by human vs. chimpanzee FOXP2 in the neuronal cell line were also differentially expressed in normal intact brains. These data demonstrate that changes in the FOXP2 gene that occurred during human evolution (subsequent to our divergence from chimpanzees) significantly affect gene expression in the human brain, potentially underlying the obligatory nature of human symbolic abilities and language in comparison to the rudimentary abilities and opportunistic expression of the homologous skills in chimpanzees. In parallel, other investigators continue to examine the relationship between FOXP2 and clinical speech and language disorders. Results confirm the hypothesis that the differences between chimpanzees and humans derive less from DNA sequence (i.e., amino acid substitutions) than from differences in gene expression and regulation.

The study by Konopka et al. (2009) provides an informative example of the unique insights that access to captive chimpanzee phenotypes, genotypes, and tissue can provide on the gamut of research from comparative genomics to behavior and biomedical. No living animals were required for this study, but it did require the following:

- Well-defined genetic sequences from both humans and chimpanzees, which of course requires access to DNA from both species as well as relatively complete information concerning the type and source of genetic variation in each species;
- Access to high-quality RNA samples from fresh chimpanzee brains, which can then be compared with similar RNA samples from human brains; and
- Detailed information about the behavioral and cognitive capacities of chimpanzees.

This type of study fulfills the general requirement to produce fundamental knowledge. Moreover, it is clear that this type of study is only possible because of the close phylogenetic relationship between humans and chimpanzees, indicating that the material from the chimpanzees provides unique, otherwise unattainable information that not only elucidates the origin of human symbolic communication, but sheds light on the mechanisms that may contribute to developmental abnormalities in this domain.

Criteria 2: All Experiments are Performed on Acquiescent Animals and in a Manner that Minimizes Distress

Inasmuch as the study required access to resources that were originally collected from living animals (genetic material, behavioral and cognitive phenotypes) or that required animals to have been alive (brain tissue harvested appropriately from deceased animals for reasons unrelated to the study), the general criteria for species-appropriate housing, acquiescence to procedures, and minimal distress for manipulations would have to be fulfilled by all animals while still alive. Importantly, maintenance in complex, species-appropriate environments would be particularly important for this type of investigation in order to maintain the species-typical ("normal") pattern of gene expression and gene regulation across the lifespan. Only in such circumstances is it likely that the derived brain and ancillary tissue will comprise the appropriate controls for the human tissues used in the same studies.

Finding

Given the information provided in the publication regarding the collection of material, the chimpanzee study used as this case example meets the committee's criteria regarding unattainable insight, acquiescence, and the minimization of pain and distress. Other examples of the application of genomic tools to behavioral or neurobehavioral investigations include the collection of tissues (including blood) that can be sequenced to provide transcriptomes (gene expression profiles). These transcriptomes, in turn, could be studied in relation to rearing experiences, temperamental characteristics, neurobehavioral traits, and other biobehavioral phenotypes to help characterize the relationship between the chimpanzee genome and the life histories of individual animals. Each such study would have to be assessed to determine whether it meets the proposed criteria.

Altruism

Studying behavior of individual chimpanzees as well as groups of animals provides insights into human behavior, thus informing scientific understanding about the nature of humans. Through their investigation of specific aspects of behavior, scientists are identifying characteristics in chimpanzees that were once thought to be unique to humans, such as certain facets of intelligence and communication, and patterns of social relationships.

Altruism is among the most contentious areas in human behavioral research. Although it has long been observed that people help each other, sometimes at great risk to their own health or well-being, there is little agreement concerning the evolutionary origin of this behavior and whether it can be expected to occur in the absence of clear self-interest

(Okasha, 2010). It is argued that studies of chimpanzees may be particularly relevant for addressing complex behaviors such as altruism because of our shared evolutionary history and recent common ancestry. Even for chimpanzees, however, disagreement remains over the degree to which the animals are sensitive to the needs of conspecifics (Horner et al., 2011).

Chosen here for a case evaluation is one study that attempts to determine whether chimpanzees actively choose to help others, and whether such help is spontaneous—and thus could be interpreted as reflecting sensitivity to the needs of another animal—or is triggered by a solicitation from the partner (Horner et al., 2011). The study design involves allowing chimpanzees to choose between two different tokens, a "selfish" token that provides a reward for the actor only, and a "prosocial" token that rewards both the actor and a partner. Seven females were tested, each with three different partners. The actors demonstrated a significant overall bias for the prosocial token, but more so when the partner either showed no reaction or engaged in neutral attention getting (not directed at the actor); attempts to pressure the actor resulted in reduced prosocial choice (Horner et al., 2011).

Criteria 1: Studies Provide Otherwise Unattainable Insight

The research objectives of the study question addressed clearly fall within the broad NIH mission to "seek fundamental knowledge about the nature and behavior of living systems" (NIH, 2011) and more specifically, meets the first criteria: to provide "otherwise unattainable insight into evolution, normal and abnormal behavior, mental health, emotion, or cognition." The insights derive from the arguably close genetic relationship between chimpanzees and human beings and the substantial similarity in brain structure and complexity—a similarity that is much less pronounced in monkeys (Sakai et al., 2011). The less complex brain structures in other old worlds of monkeys result in a tendency toward a simpler pattern of signaling and signal recognition. It was of particular interest that, in this study, actors behaved prosocially toward their partners irrespective of relative social status, genetic relationship, or expectation of reciprocity. These results imply that human beings may have a tendency to help other individuals unconditionally, at least when the help can be given at no cost.

Criteria 2: All Experiments Are Performed on Acquiescent Animals and in a Manner That Minimizes Distress

In assessing the degree of acquiescence and distress on the part of the subjects, the study reported that the seven adult female chimpanzees "volunteered to participate and were willing to exchange tokens with an investigator." However, no details regarding the definition of "volunteer" were given in the manuscript or the associated methodology. The study did describe the conditions under which the animals were maintained. Specifically, they were housed in a large outdoor grass enclosure with climbing structures as part of a long-established social group comprising 11 females and one male. There were two associated buildings, one with indoor sleeping quarters and a second building designed for cognitive research testing. No details were provided on the dimensions of these buildings.

Finding

This study involved temporary removal of animals from their usual housing and social group to engage in a cognitive task paired with other chimpanzees. The information provided suggests that chimpanzee use in this study could meet all criteria if more complete descriptions of the handling and housing were provided. Specifically, the investigators would have to substantiate the statement that the animals "volunteered" for the procedures, confirm that the indoor sleeping quarters were of sufficient size for the species, and demonstrate that the cognitive testing apparatus would meet all enrichment requirements for this species.

This study exemplifies the numerous cognitive investigations that have been done in chimpanzees. As a group, these studies have demonstrated that human cognitive abilities with respect to the manipulation of the social environment extend to chimpanzees in a variety of domains that include altruism, deception, and grief. Many such studies would be similarly approvable under the proposed guidelines; in other instances they might be limited if, for example, they provided unattainable insights but did not meet the need for acquiescence and minimization of distress.

Cognition

Joint attention occurs when one animal alerts another to the presence of a stimulus by means of gestural or vocal communication. It is generally thought that a breakdown in the ability to initiate joint attention may be a predictor for autism spectrum disorders or other neurodevelopmental disorders (Mundy et al., 2010). Unfortunately, knowledge of developmental mechanisms of joint attention are poorly understood because some functional imaging techniques used in adults are difficult to administer to children while others, like positron emission tomography (PET), cannot be administered without some risk to normal young subjects. The chimpanzee has been used as a model organism to study the neurodevelopmental basis of joint attention and similar phenomena and to increase the understanding of the development trajectory of human communicative phenomena. This is because, in humans, joint attention (including both gestures and vocalization) is associated with hemispheric lateralization, particularly in the portion of the inferior frontal gyrus (IFG), termed Broca's area, and is thought to have evolved from a lateralized manual communication system present in the common ancestor of humans and chimpanzees (Corballis, 2002; Kingstone et al., 2000). Chosen here for case evaluation is a PET study designed to determine whether chimpanzees possess a gesture and vocalization-activated brain region homologous with the IFG (Broca's area), which in humans is most often enlarged on the left side to indicate significant left-lateralized patterns of activation during communication (Taglialatela et al., 2008). While it has been previously shown that chimpanzees engage in joint communication and exhibit structural asymmetry in the brain, it had not been demonstrated that the brain regions underlying joint attention were the same as those underlying homologous communicative behaviors in humans. In particular, it was not known whether the IFG and related cortical and subcortical areas were preferentially activated during gestural or vocal communication in the chimpanzee. The presence of similarly activated underlying brain structures would suggest that chimpanzees could be used to model human communication development.

Criteria 1: Studies Provide Otherwise Unattainable Insight

The forgoing description suggests that the study does fulfill the need to provide fundamental knowledge gain. Moreover, because the chimpanzee and humans both uniquely share a highly convoluted and lateralized cerebral cortex and the ability to engage in joint attention, it is likely to provide otherwise unattainable insight into the neurodevelopment of communication and, by implication, communicative disorders. Furthermore, while the modern imaging modalities necessary to map neurodevelopment can be applied to both chimpanzees and adult human beings, the application of these techniques to children is often limited by logistical and ethical considerations. Finally, even if all imaging techniques (even those involving unacceptably high radiation exposure) could be applied to children, the study of neurodevelopment would be arguably enhanced by the availability of a comparative model like the chimpanzee.

This study provided the first direct evidence that the neuroanatomical structures underlying communicative signals in chimpanzees are homologous with those present in humans. Furthermore, chimpanzees engaging in communicative gestures—like human beings—activated the left IFG (Broca's area) in conjunction with other cortical and subcortical brain areas, providing strong evidence in support of the hypothesis that the neurological substrates underlying language production in the human brain were present in the last common ancestor of humans and chimpanzees.

Criteria 2: All Experiments Are Performed on Acquiescent Animals and in a Manner That Minimizes Distress

Chimpanzees were initially separated from groupmates, but maintained within their home enclosure. The animals were then provided with a sweetened drink that contained the radioligand ^{18}F-fluorodeoxyglucose (^{18}F-FDG), which initiated a 40-minute uptake period, during which the radioligand would bind to parts of the brain that were being activated by the chimpanzee's behavior. During the uptake period, the investigator sitting outside the enclosure placed a favored food item just beyond reach, a situation that predictably elicited both gestural and vocal signaling from subjects. Experimenters would periodically respond to subjects with both vocal communication and small food rewards. At the end of the 40-minute uptake period, chimpanzees received an intramuscular injection of an anesthetic agent and were transported to the PET facility. All animals had been previously trained with positive reinforcement to present for such anesthesia. Following scanning, animals were allowed to fully recover before being reintroduced to their social groups. It should be noted that three chimpanzees were used in the study, and each animal was scanned on two occasions—once following the gesture and vocalization task and separately following a control task that did not require communicative interaction.

Review of the study indicates that it was conducted in a manner consistent with acquiescence. Animals voluntarily engaged in behavioral testing during the awake part of the procedure (i.e., uptake) and, furthermore, were trained to present for anesthesia. As described in the study, anesthesia persisted for about 50 minutes (transport to and from the PET facility, PET scanning). Animals were allowed to recover (and radioligands allowed to decay) for approximately 18 hours, after which they were returned to their social group. It is important to note that this study likely exposed the animals to more than "minimal" distress in that animals were fasted for 5 hours prior to the procedure, sedated for at least 50 minutes, and then were separated from their social groups to allow recovery. The total time required for the manipulation was probably 28-32 hours. However, the effects of this amount of separation from the social group and handling must be judged against the unique contribution made by the study and the small number of acquiescent animals involved. It should be noted that a complete veterinary examination would involve a similar if not longer period of fasting, social group separation, and anesthesia.

Finding

In view of the scientific benefits compared to the temporary negative impacts on the animal subjects (separation and anesthesia), this study could potentially meet all criteria for approval if sufficient additional assurance were provided that the animals were maintained in species-appropriate housing and groupings and that the number and duration of procedures imposed on individual animals was minimized in a manner consistent with criteria described earlier in this report.

FUTURE USE OF CHIMPANZEES IN BIOMEDICAL AND BEHAVIORAL RESEARCH

As highlighted throughout this report, over many years scientific advances that have led directly to the development of preventive and therapeutic products for life-threatening or debilitating diseases and disorders have been dependent on scientific knowledge obtained through experiments using the chimpanzee. In addition, many preliminary proof-of-concept experiments have been carried out in the chimpanzee; for example, development of human and humanized monoclonal antibody therapies have required preclinical testing in the chimpanzee (Iwarson et al., 1985). The same has been the case for early evaluation of therapeutic concepts based on RNAi, microRNA, and antisense RNA (e.g., for treating chronic HCV infection), and for evaluation of TLR7 antagonists (e.g., for treating chronic HBV infection) (Lanford et al., 2011).

The National Institute of Allergy and Infectious Diseases at the NIH has identified eight instances over the past two decades where research on new (or newly recognized), emerging, and reemerging infectious diseases has called for use of the chimpanzee to answer crucial questions pertaining to pathogenesis, prevention, control, or therapy. In five of these, the chimpanzee is still being used.[11] At the same time, as has been the case rather often in the past, an important new, emerging, or reemerging disease may present treatment, prevention, and/or control problems that defy available alternative experimental approaches, including the most novel, innovative approaches, and therefore may require use of the chimpanzee—rare as this may be, this possibility cannot be discounted over the long term. The committee recognizes that the limited number of available animals and the potential need to perform experiments under conditions of biocontainment could potentially constrain the value of the chimpanzee during a public health emergency. The similarity in the neuroanatomy between the human and the chimpanzee may make it a model for neuropsychiatric disorders, for example, expressing human risk genes via viral vectors or from optogenetic methods that exploit the chimpanzee functional neuroanotomy.

However, in this case the past is not necessarily prelude—great progress is being made in developing alternatives to the chimpanzee; more studies are using other non-human primates (Ben-Yehudah et al., 2010; Couto and Kolykhalov, 2006; Pan et al., 2010; Suomi, 2006), genetically modified (knock-out, knock-in) mice (Chen et al., 2011a; de Jong et al., 2010; Dorner et al., 2011; Kneteman and Mercer, 2005; Lindenbach et al., 2005; Ma et al., 2010; Ploss and Rice, 2009), and even in silico technologies (Hosea, 2011; Qiu et al., 2011; Valerio, 2011). In some instances, "preclinical studies" in humans, that is, expanded studies carried out

---

[11]Encephalitozoon cuniculi; Helicobacter pylori; Hepatitis C (ongoing); Hepatitis E (ongoing); Human herpesvirus 8 (ongoing); Human herpesvirus 6 (ongoing); Streptococcus, Group A; Staphylococcus aureus (ongoing) (NIAID, 2011).

in the field during disease outbreaks, have served as an alternative to the use of the chimpanzee.

Finding

The committee cannot predict or forecast future need of the chimpanzee animal model and encourages use of the criteria established in this report when assessing the potential necessity of chimpanzees for future research uses.

CONCLUSIONS AND RECOMMENDATIONS

Animal models serve as a critical research tool in facilitating the advancement of the public's health. The chimpanzee's genetic proximity to humans and the resulting biological and behavioral characteristics not only make it a uniquely valuable species for certain types of research, but also demand a greater justification for conducting research using this animal model. As this report demonstrates, the committee's conclusions and recommendations are predicated on the advances that have been made by the scientific community in developing and using alternative models to the chimpanzee, such as studies involving human subjects, other non-human primates, genetically modified mice, in vitro systems, and in silico technologies. Having reviewed and analyzed contemporary and anticipated biomedical and behavioral research, the committee offers the following three conclusions and two recommendations.

Conclusion 1: Assessing the Necessity of the Chimpanzee for Biomedical Research

Having explored and analyzed contemporary and anticipated biomedical research questions, the committee concludes:
- The chimpanzee has been a valuable animal model.
- Based on a set of principles that ensure ethical treatment of chimpanzees, the committee established criteria to determine the necessity for the use of chimpanzees in current biomedical or behavioral research.
- While the chimpanzee has been a valuable animal model in past research, most current use of chimpanzees for biomedical re-

search is unnecessary, based on the criteria established by the committee, except potentially for two current research uses:

- o Development of future monoclonal antibody therapies will not require the chimpanzee, due to currently available technologies. However, there may be a limited number of monoclonal antibodies already in the developmental pipeline that may require the continued use of chimpanzees.
- o The committee was evenly split and unable to reach consensus on the necessity of the chimpanzee for the development of a prophylactic HCV vaccine. Specifically, the committee could not reach agreement on whether a preclinical challenge study using the chimpanzee model was necessary and if or how much the chimpanzee model would accelerate or improve prophylactic HCV vaccine development.

- The present trajectory indicates a decreasing scientific need for chimpanzee studies due to the emergence of non-chimpanzee models and technologies.
- Development of non-chimpanzee models requires continued support by the NIH.
- A new, emerging, or reemerging disease or disorder may present challenges to treatment, prevention, and/or control that defy non-chimpanzee models and technologies and therefore may require the future use of the chimpanzee.
- Application of the committee's criteria would provide a framework to assess scientific necessity to guide the future use of chimpanzees in biomedical research.

Recommendation 1: The National Institutes of Health should limit the use of chimpanzees in biomedical research to those studies that meet the following three criteria:

1. There is no other suitable model available, such as in vitro, non-human in vivo, or other models, for the research in question; and
2. The research in question cannot be performed ethically on human subjects; and

3. Forgoing the use of chimpanzees for the research in question will significantly slow or prevent important advancements to prevent, control, and/or treat life-threatening or debilitating conditions.

Animals used in the proposed research must be maintained either in ethologically appropriate physical and social environments or in natural habitats. Biomedical research using stored samples is exempt from these criteria.

Conclusion 2: Assessing the Necessity of the Chimpanzee for Comparative Genomics Research

Having reviewed comparative genomics research, the committee concludes the chimpanzee may be necessary for understanding human development, disease mechanisms, and susceptibility because of the genetic proximity of the chimpanzee to humans. Furthermore, comparative genomics research poses minimal risk of pain and distress to the chimpanzee in instances where samples are collected from living animals and poses no risk when biological materials are derived from existing samples. Application of the committee's criteria would provide a framework to assess scientific necessity to guide the future use of chimpanzees in comparative genomics research that requires samples collected from living animals.

Conclusion 3: Assessing the Necessity of the Chimpanzee for Behavioral Research

Having explored and analyzed contemporary and anticipated behavioral research questions, the committee concludes that chimpanzees may be necessary for obtaining otherwise unattainable insights to support understanding of social, neurological, and behavioral factors that include the development, prevention, or treatment of disease. Application of the committee's criteria would provide a framework to assess scientific necessity to guide the future use of chimpanzees in behavioral research.

Recommendation 2: The National Institutes of Health should limit the use of chimpanzees in comparative genomics and behavioral research to those studies that meet the following two criteria:

1. Studies provide otherwise unattainable insight into comparative genomics, normal and abnormal behavior, mental health, emotion, or cognition; and
2. All experiments are performed on acquiescent animals, using techniques that are minimally invasive, and in a manner that minimizes pain and distress.

Animals used in the proposed research must be maintained either in ethologically appropriate physical and social environments or in natural habitats. Comparative genomics and behavioral research using stored samples are exempt from these criteria.

The criteria set forth in the report are intended to guide not only current research policy, but also decisions regarding potential use of the chimpanzee model for future research. The committee acknowledges that imposing an outright and immediate prohibition of funding could cause unacceptable losses to research programs as well as have an impact on the animals. Therefore, although the committee was not asked to consider how its recommended policies should be implemented, it believes that the NIH should evaluate the necessity of the chimpanzee in all grant renewals and future research projects using the chimpanzee model based on the committee's criteria.

In March 1989 the NIH chartered the Interagency Animal Model Committee (IAMC) "to provide oversight of all federally supported biomedical and behavioral research involving chimpanzees" (NIH, unpublished). As indicated in its charter:

> The IAMC review mechanism represents a commitment to the public and the U.S. Congress to promote the conservation and care of chimpanzees when this species is the best or possibly the only model for the conduct of research to advance scientific knowledge and to address questions that have significant impact on public health.
>
> The IAMC shall review all federally-supported research protocols involving the use of chimpanzees before the initiation of the study. Prior to this review, the

project must be reviewed and approved by intramural scientific program staff or an extramural initial review group and by the appropriate Animal Care and Use Committee. The IAMC's evaluation constitutes an additional level of scientific review, focusing on such factors as the appropriateness of the animal model, appropriateness of the numbers of animals, the availability of the animals, the degree of invasiveness of the procedures, and any unnecessary duplication. (NIH, unpublished)

Appointment of the IAMC is evidence that the NIH has determined that the conservation and care of chimpanzees requires additional oversight. Membership on the Interagency Animal Model Committee is restricted to federal employees from the Department of Health and Human Services (including the NIH, CDC, and FDA), Department of Veterans Affairs, and Department of Defense. The committee believes that assessment of potential future use of the chimpanzee would be strengthened and the process made more credible by establishing an independent oversight committee that uses the recommended criteria and includes public representatives as well as individuals with scientific expertise, both in the use of chimpanzees and alternative models, in areas of research that have the potential for chimpanzee use.

A

References

Abee, C. 2011a. (Michale E. Keeling Center for Comparative Medicine and Research of the University of Texas MD Anderson Cancer Center, Bastrop, TX). Letter to Jeffrey Kahn: Chimpanzee mAb studies with attachments: Importance of chimpanzees in mAb research.pdf, Responses to IOM committee questions on mAb research with chimpanzees.pdf. September 30, 2011.

Abee, C. 2011b. (Michale E. Keeling Center for Comparative Medicine and Research of the University of Texas MD Anderson Cancer Center, Bastrop, TX). Email to authors: Chimpanzee studies and provision of research resources with attachments: Keeling Center from the University of Texas, KCCMR: Chimpanzee studies by year and type.pdf. July 1, 2011.

Abee, C. 2011c. (Michale E. Keeling Center for Comparative Medicine and Research of the University of Texas MD Anderson Cancer Center, Bastrop, TX). Email to authors: IOM chimp study—verification of numbers and ages of chimpanzees at the Michale E. Keeling Center for Comparative Medicine and Research. May 17, 2011.

Abee, C., M. Lammey, T. Rowell, J. VandeBerg, and S. Zola. 2011. Conference call with the directors of the Yerkes, New Iberia, Alamogordo, Southwest, and MD Anderson chimpanzee centers, Washington, DC, July 1, 2011.

Alnylam Pharmaceuticals. 2011. Phase 2b study of aln-rsv01 in lung transplant patients infected with respiratory syncytial virus (RSV). *ClinicalTrials.gov: A service of the U.S. National Institutes of Health.* http://clinicaltrials.gov/ct2/show/NCT01065935?term=nct01065935&rank=1NLMIdentifier:NCT01065935 (accessed August 24, 2011).

Altaweel, L., Z. Chen, M. Moayeri, X. Cui, Y. Li, J. Su, Y. Fitz, S. Johnson, S. H. Leppla, R. Purcell, and P. Q. Eichacker. 2011. Delayed treatment with w1-mab, a chimpanzee-derived monoclonal antibody against protective antigen, reduces mortality from challenges with anthrax edema or lethal toxin in rats and with anthrax spores in mice. *Crit Care Med* 39(6):1439-1447.

An, Z. 2010. Monoclonal antibodies—a proven and rapidly expanding therapeutic modality for human diseases. *Protein Cell* 1(4):319-330.

Animal Welfare Act (New Zealand). 1999. Public Act 1999, no. 142. Date of Assent 14 October 1999.

Australian Government National Health and Medical Research Council. 2003. *Policy on the use of non-human primates for scientific purposes.* http://www.nhmrc.gov.au/_files_nhmrc/publications/attachments/ea14.pdf (accessed October 213, 2011).

AZA (Association of Zoos and Aquariums) Ape TAG. 2010. *Chimpanzee (Pan troglodytes) care manual.* Silver Spring, MD: Association of Zoos and Aquariums.

Bailey, J. 2008. An assessment of the role of chimpanzees in AIDS vaccine research. *ALTA* 36(4):381-428.

Bailey, J. 2010a. An assessment of the use of chimpanzees in hepatitis C research past, present and future: Validity of the chimpanzee model. *ALTA* 38(5):387-418.

Bailey, J. 2010b. An assessment of the use of chimpanzees in hepatitis C research past, present and future: 2. Alternative replacement methods. *ALTA* 38(6):471-494.

Bassett, S. E., K. M. Brasky, and R. E. Lanford. 1998. Analysis of hepatitis C virus-inoculated chimpanzees reveals unexpected clinical profiles. *J Virol* 72(4):2589-2599.

Bassett, S. E., D. L. Thomas, K. M. Brasky, and R. E. Lanford. 1999. Viral persistence, antibody to e1 and e2, and hypervariable region 1 sequence stability in hepatitis C virus-inoculated chimpanzees. *J Virol* 73(2):1118-1126.

Bateson, P. 2011. *Review of research using non-human primates: Report of a panel chaired by Professor Sir Patrick Bateson FRS.* Medical Research Council (MRC). London, England.

Beck, A., T. Wurch, C. Bailly, and N. Corvaia. 2010. Strategies and challenges for the next generation of therapeutic antibodies. *Nat Rev Immunol* 10(5):345-352.

Becker, P. D., N. Legrand, C. M. van Geelen, M. Noerder, N. D. Huntington, A. Lim, E. Yasuda, S. A. Diehl, F. A. Scheeren, M. Ott, K. Weijer, H. Wedemeyer, J. P. Di Santo, T. Beaumont, C. A. Guzman, and H. Spits. 2010. Generation of human antigen-specific monoclonal IgM antibodies using vaccinated "human immune system" mice. *PLoS One* 5(10).

Bem, R. A., J. B. Domachowske, and H. F. Rosenberg. 2011. Animal models of human respiratory syncytial virus disease. *Am J Physiol Lung Cell Mol Physiol* 301(2):L148-L156.

Ben-Yehudah, A., C. A. T. Easley, B. P. Hermann, C. Castro, C. Simerly, K. E. Orwig, S. Mitalipov, and G. Schatten. 2010. Systems biology discoveries using non-human primate pluripotent stem and germ cells: Novel gene and genomic imprinting interactions as well as unique expression patterns. *Stem Cell Res Ther* 1(3):24.

Bennett, B. T., C. R. Abee, and R. Henrickson. 1995. *Nonhuman primates in biomedical research.* 2 vols., *American College of Laboratory Animal Medicine Series.* San Diego, CA: Academic Press.

Bettauer, R. H. 2010. Chimpanzees in hepatitis C virus research: 1998-2007. *J Med Primatol* 39(1):9-23.

Bettauer, R. H. 2011. Systematic review of chimpanzee use in monoclonal antibody research and drug development: 1981-2010. *ALTEX* 28(2):103-116.

Bissig, K. D., S. F. Wieland, P. Tran, M. Isogawa, T. T. Le, F. V. Chisari, and I. M. Verma. 2010. Human liver chimeric mice provide a model for hepatitis B and C virus infection and treatment. *J Clin Invest* 120(3):924-930.

Blanco, J. C., M. S. Boukhvalova, K. A. Shirey, G. A. Prince, and S. N. Vogel. 2010. New insights for development of a safe and protective RSV vaccine. *Hum Vaccin* 6(6):482-492.

Boonstra, A., L. J. van der Laan, T. Vanwolleghem, and H. L. Janssen. 2009. Experimental models for hepatitis C viral infection. *Hepatology* 50(5):1646-1655.

BPRC (Biomedical Primate Research Centre). 2011. *Animal welfare: Relocation of chimpanzees.* http://www.bprc.nl/BPRCE/L3/RelocChimps.html (accessed September 30, 2011).

Bukh, J., X. Forns, S. U. Emerson, and R. H. Purcell. 2001. Studies of hepatitis C virus in chimpanzees and their importance for vaccine development. *Intervirology* 44(2-3):132-142.

Byrd, L. G., and G. A. Prince. 1997. Animal models of respiratory syncytial virus infection. *CID* 25:1363-1368.

Carroll, S. S., S. Ludmerer, L. Handt, K. Koeplinger, N. Y. R. Zhang, D. Graham, M. E. Davies, M. MacCoss, D. Hazuda, and D. B. Olsen. 2009. Robust antiviral efficacy upon administration of a nucleoside analog to hepatitis C virus-infected chimpanzees. *Antimicrob Agents Chemother* 53(3):926-934.

Catanese, M. T., H. Ansuini, R. Graziani, T. Huby, M. Moreau, J. K. Ball, G. Paonessa, C. M. Rice, R. Cortese, A. Vitelli, and A. Nicosia. 2010. Role of scavenger receptor class B type in hepatitis C virus entry: Kinetics and molecular determinants. *J Virol* 84(1):34-43.

CBER (Center for Biologics Evaluation and Research) and CDER (Center for Drug Evaluation and Research). 2009. *Guidance for industry.* Rockville, MD: U.S. Department of Health and Human Services, Food and Drug Administration.

Chapman, K., N. Pullen, M. Graham, and I. Ragan. 2007. Preclinical safety testing of monoclonal antibodies: The significance of species relevance. *Nat Rev Drug Discov* 6(2):120-126.

Chapman, K., N. Pullen, L. Coney, M. Dempster, L. Andrews, J. Bajramovic, P. Baldrick, L. Buckley, A. Jacobs, G. Hale, C. Green, I. Ragan, and V. Robinson. 2009. Preclinical development of monoclonal antibodies: Considerations for the use of non-human primates. *MAbs* 1(5):505-516.

Chen, Z., P. Earl, J. Americo, I. Damon, S. K. Smith, Y. H. Zhou, F. Yu, A. Sebrell, S. Emerson, G. Cohen, R. J. Eisenberg, J. Svitel, P. Schuck, W. Satterfield, B. Moss, and R. Purcell. 2006a. Chimpanzee/human mAbs to vaccinia virus b5 protein neutralize vaccinia and smallpox viruses and protect mice against vaccinia virus. *Proc Natl Acad Sci USA* 103(6):1882-1887.

Chen, Z., M. Moayeri, Y. H. Zhou, S. Leppla, S. Emerson, A. Sebrell, F. Yu, J. Svitel, P. Schuck, M. St. Claire, and R. Purcell. 2006b. Efficient neutralization of anthrax toxin by chimpanzee monoclonal antibodies against protective antigen. *J Infect Dis* 193(5):625-633.

Chen, C. M., Y. He, L. Lu, H. B. Lim, R. L. Tripathi, T. Middleton, L. E. Hernandez, D. W. Beno, M. A. Long, W. M. Kati, T. D. Bosse, D. P. Larson, R. Wagner, R. E. Lanford, W. E. Kohlbrenner, D. J. Kempf, T. J. Pilot-Matias, and A. Molla. 2007a. Activity of a potent hepatitis C virus polymerase inhibitor in the chimpanzee model. *Antimicrob Agents Chemother* 51(12):4290-4296.

Chen, Z., P. Earl, J. Americo, I. Damon, S. K. Smith, F. Yu, A. Sebrell, S. Emerson, G. Cohen, R. J. Eisenberg, I. Gorshkova, P. Schuck, W. Satterfield, B. Moss, and R. Purcell. 2007b. Characterization of chimpanzee/human monoclonal antibodies to vaccinia virus a33 glycoprotein and its variola virus homolog in vitro and in a vaccinia virus mouse protection model. *J Virol* 81(17):8989-8995.

Chen, Z., M. Moayeri, D. Crown, S. Emerson, I. Gorshkova, P. Schuck, S. H. Leppla, and R. H. Purcell. 2009. Novel chimpanzee/human monoclonal antibodies that neutralize anthrax lethal factor, and evidence for possible synergy with anti-protective antigen antibody. *Infect Immun* 77(9):3902-3908.

Chen, A. A., D. K. Thomas, L. L. Ong, R. E. Schwartz, T. R. Golub, and S. N. Bhatia. 2011a. Humanized mice with ectopic artificial liver tissues. *Proc Natl Acad Sci USA* 108(29):11842-11847.

Chen, Z., K. Chumakov, E. Dragunsky, D. Kouiavskaia, M. Makiya, A. Neverov, G. Rezapkin, A. Sebrell, and R. Purcell. 2011b. Chimpanzee-human monoclonal antibodies for treatment of chronic poliovirus excretors and emergency postexposure prophylaxis. *J Virol* 85(9):4354-4362.

Chen, Z., R. Schneerson, J. Lovchik, C. R. Lyons, H. Zhao, Z. Dai, J. Kubler-Kielb, S. H. Leppla, and R. H. Purcell. 2011c. Pre- and postexposure protection against virulent anthrax infection in mice by humanized monoclonal antibodies to bacillus anthracis capsule. *Proc Natl Acad Sci USA* 108(2):739-744.

Chin, J., R. L. Magoffin, L. A. Shearer, J. H. Schieble, and E. H. Lennette. 1969. Field evaluation of a respiratory syncytial virus vaccine and a trivalent parainfluenza virus vaccine in a pediatric population. *Am J Epidemiol* 89(4):449-463.

Choo, Q. L., G. Kuo, A. J. Weiner, L. R. Overby, D. W. Bradley, and M. Houghton. 1989. Isolation of a cDNA clone derived from a blood-borne non-A, non-B viral hepatitis genome. *Science* 244(4902):359-362.

Choo, Q. L., G. Kuo, R. Ralston, A. Weiner, D. Chien, G. Van Nest, J. Han, K. Berger, K. Thudium, C. Kuo et al. 1994. Vaccination of chimpanzees against infection by the hepatitis C virus. *Proc Natl Acad Sci USA* 91(4):1294-1298.

Cohen, J. 2007a. The endangered lab chimp. *Science* 315(5811):450-452.

Cohen, J. 2007b. NIH to end chimp breeding for research. *Science 316(5829):*1265.

CHMP (Committee for Medicinal Products for Human Use). 2011. *CHMP: Overview.* http://www.ema.europa.eu/ema/index.jsp?curl=pages/about_us/general/general_content_000095.jsp&murl=menus/about_us/about_us.jsp&mid=WC0b01ac0580028c7a (accessed September 16, 2011).

Congress of Spain, IX Legislature. 2008. *Official bulletin of the general courts* (No. 19). http://www.congreso.es/public_oficiales/L9/CONG/BOCG/D/D_019.PDF (accessed August 23, 2011).

Cooper, S., A. L. Erickson, E. J. Adams, J. Kansopon, A. J. Weiner, D. Y. Chien, M. Houghton, P. Parham, and C. M. Walker. 1999. Analysis of a successful immune response against hepatitis C virus. *Immunity* 10(4):439-449.

Corballis, M. C. 2002. *From hand to mouth: The origins of language.* Princeton, NJ: Princeton University Press.

Council of Europe. 2006. *Appendix A of the European convention for the protection of vertebrate animals used for experimental and other scientific purposes (ETS No. 123) guidelinges for accomodation and care of animals (article 5 of the convention) approved by the multilateral consulation.* Council of Europe. Strasbourg.

Couto, L. B., and A. A. Kolykhalov. 2006. Animal models for HCV study. In *Hepatitis C viruses: Genomes and molecular biology,* edited by S. L. Tan. Wymondham, Norfolk, UK: Horizon Bioscience.

Dahari, H., S. M. Feinstone, and M. E. Major. 2010. Meta-analysis of hepatitis C virus vaccine efficacy in chimpanzees indicates an importance for structural proteins. *Gastroenterology* 139(3):965-974.

de Jong, Y. P., C. M. Rice, and A. Ploss. 2010. New horizons for studying human hepatotropic infections. *J Clin Invest* 120(3):650-653.

de Marco, A. 2011. Biotechnological applications of recombinant single-domain antibody fragments. *Microb Cell Fact* 10:44.

De Vos, R., C. Verslype, E. Depla, J. Fevery, B. Van Damme, V. Desmet, and T. Roskams. 2002. Ultrastructural visualization of hepatitis C virus components in human and primate liver biopsies. *J Hepatol* 37(3):370-379.

Demarest, S. J., and S. M. Glaser. 2008. Antibody therapeutics, antibody engineering, and the merits of protein stability. *Curr Opin Drug Discov Devel* 11(5):675-687.

DeVincenzo, J., R. Lambkin-Williams, T. Wilkinson, J. Cehelsky, S. Nochur, E. Walsh, R. Meyers, J. Gollob, and A. Vaishnaw. 2010. A randomized, double-blind, placebo-controlled study of an RNAi-based therapy directed against respiratory syncytial virus. *Proc Natl Acad Sci USA* 107(19):8800-8805.

DHS (U.S. Department of Homeland Security). 2007. *Care and use of animals in research.*. Washington, DC: U.S. Government, Management Directives System.

Domachowske, J. B., C. A. Bonville, and H. F. Rosenberg. 2004. Animal models for studying respiratory syncytial virus infection and its long term effects on lung function. *Pediatr Infect Dis J* 23(Suppl):S228-S234.

Dorner, M., J. A. Horwitz, J. B. Robbins, W. T. Barry, Q. Feng, K. Mu, C. T. Jones, J. W. Schoggins, M. T. Catanese, D. R. Burton, M. Law, C. M. Rice, and A. Ploss. 2011. A genetically humanized mouse model for hepatitis C virus infection. *Nature* 474(7350):208-211.

Eastwood, D., L. Findlay, S. Poole, C. Bird, M. Wadhwa, M. Moore, C. Burns, R. Thorpe, and R. Stebbings. 2010. Monoclonal antibody tgn1412 trial failure explained by species differences in cd28 expression on cd4+ effector memory T-cells. *Br J Pharmacol* 161(3):512-526.

Ehrlich, P. H., Z. A. Moustafa, K. E. Harfeldt, C. Isaacson, and L. Ostberg. 1990. Potential of primate monoclonal antibodies to substitute for human antibodies: Nucleotide sequence of chimpanzee fab fragments. *Hum Antibodies Hybridomas* 1(1):23-26.

Else, J. G. 2011. (Yerkes National Primate Research Center, Atlanta, GA). Email to authors: Chimpanzee ages with file attachment: yerkeschimpagereport.pdf. October 31, 2011.

EMEA (European Medicines Agency). 2008. *ICH topic m3 (r2) non-clinical safety studies for the conduct of human clinical trials and marketing authorization for pharmaceuticals.* European Medincines Agency. London, England.

EMEA. 2011a. *ICH guideline S6 (R1)—preclinical safety evaluation of biotechnology-derived pharmaceuticals.* EMA/CHMP/ICH/731268/1998 European Medicines Agency.

EMEA. 2011b. *CHMP assessment report, victrelis.* http://www.ema.europa.eu/docs/en_GB/document_library/EPAR_-_Public_assessment_report/human/002332/WC500109789.pdf (accessed September 13, 2011).

Enard, W., M. Przeworski, S. E. Fisher, C. S. Lai, V. Wiebe, T. Kitano, A. P. Monaco, and S. Paabo. 2002. Molecular evolution of FOXP2, a gene involved in speech and language. *Nature* 418(6900):869-872.

Erickson, A. L., Y. Kimura, S. Igarashi, J. Eichelberger, M. Houghton, J. Sidney, D. McKinney, A. Sette, A. L. Hughes, and C. M. Walker. 2001. The outcome of hepatitis C virus infection is predicted by escape mutations in epitopes targeted by cytotoxic T lymphocytes. *Immunity* 15(6):883-895.

European Communities, and Office for Official Publications. 1986. Council directive on the approximation of laws, regulations and administrative provisions of member states regarding the protection of animals used for experimental and other scientific purposes. *Official Journal of the European Communities* No. L 358(18. 12. 86):28 p.

European Parliament. 2007. *Declaration of the European Parliament on primates in scientific experiments.* European Parliament.

European Union. 2010. Directive 2010/63/EU of the European parliament and of the council of 22 September 2010 on the protection of animals used for scientific purposes. *Official Journal of the European Union* No. L 276/33 (20. 10. 2010).

FDA (U.S. Food and Drug Administration). 1986. *OKT3: Ortho pharmaceutical corporation: Summary basis of approval of muromonab-cd3.* http://www.foiservices.com/ (accessed (September 8, 2011).

FDA. 2011a. *Title 21—food and drugs—chapter 1—subchapter D—drugs for human use part 314 application for FDA approval to market a new drug* (21CFR314). http://www.accessdata.fda.gov/scripts/cdrh/cfdocs/cfcfr/cfrsearch.cfm?cfrpart=cfrsearch.cfm?cfrpart=314&showfr=1&subpartnode=21:5.0.1.1.4.9 (accessed September 20, 2011).

FDA. 2011b. *Title 21—food and drugs—chapter 1—subchapter F—biologics part 601 licensing* (21CFR601). http://www.accessdata.fda.gov/scripts/cdrh/cfdocs/cfcfr/CFRSearch.cfm?CFRPart=601&showFR=1&subpartNode=21:7.0.1.1.2.8 (accessed September 20, 2011).

Folgori, A., S. Capone, L. Ruggeri, A. Meola, E. Sporeno, B. B. Ercole, M. Pezzanera, R. Tafi, M. Arcuri, E. Fattori, A. Lahm, A. Luzzago, A. Vitelli, S. Colloca, R. Cortese, and A. Nicosia. 2006. A T-cell HCV vaccine eliciting effective immunity against heterologous

Fulginiti, V. A., J. J. Eller, O. F. Sieber, J. W. Joyner, M. Minamitani, and G. Meiklejohn. 1969. Respiratory virus immunization. A field trial of two inactivated respiratory virus vaccines; an aqueous trivalent parainfluenza virus vaccine and an alum-precipitated respiratory syncytial virus vaccine. *Am J Epidemiol* 89(4):435-448.

Galun, E., T. Burakova, M. Ketzinel, I. Lubin, E. Shezen, Y. Kahana, A. Eid, Y. Ilan, A. Rivkind, G. Pizov et al. 1995. Hepatitis C virus viremia in SCID-->BNX mouse chimera. *J Infect Dis* 172(1):25-30.

Garrone, P., A. C. Fluckiger, P. E. Mangeot, E. Gauthier, P. Dupeyrot-Lacas, J. Mancip, A. Cangialosi, I. Du Chene, R. LeGrand, I. Mangeot, D. Lavillette, B. Bellier, F. L. Cosset, F. Tangy, D. Klatzmann, and C. Dalba. 2011. A prime-boost strategy using virus-like particles pseudotyped for HCV proteins triggers broadly neutralizing antibodies in macaques. *Sci Transl Med* 3(94):94ra71.

Gewin, V. 2011. Hepatitis C mouse model a major milestone: Development paves the way for testing potential vaccines. *Nature News.* http://www.nature.com/news/2011/110608/full/news.2011.356.html (accessed October 13, 2011).

Gibbs, R. A., J. Rogers, M. G. Katze, R. Bumgarner, G. M. Weinstock, E. R. Mardis, K. A. Remington, R. L. Strausberg, J. C. Venter, R. K. Wilson, M. A. Batzer, C. D. Bustamante, E. E. Eichler, M. W. Hahn, R. C. Hardison, K. D. Makova, W. Miller, A. Milosavljevic, R. E. Palermo, A. Siepel, J. M. Sikela, T. Attaway, S. Bell, K. E. Bernard, C. J. Buhay, M. N. Chandrabose, M. Dao, C. Davis, K. D. Delehaunty, Y. Ding, H. H. Dinh, S. Dugan-Rocha, L. A. Fulton, R. A. Gabisi, T. T. Garner, J. Godfrey, A. C. Hawes, J. Hernandez, A S. Hines, M. Holder, J. Hume, S. N. Jhangiani, V. Joshi, Z. M. Khan, E. F. Kirkness, A. Cree, R. G. Fowler, S. Lee, L. R. Lewis, Z. Li, Y. S. Liu, S. M. Moore, D. Muzny, L. V. Nazareth, D. N. Ngo, G. O. Okwuonu, G. Pai, D. Parker, H. A. Paul, C. Pfannkoch, C. S. Pohl, Y. H. Rogers, S. J. Ruiz, Sabo, J. Santibanez, B. W. Schneider, S. M. Smith, E. Sodergren, A. F. Svatek, T. R. Utterback, S. Vattathil, W. Warren, C. S. White, A. T. Chinwalla, Y. Feng, A. L. Halpern, L. W. Hillier, X. Huang, P. Minx, J. O. Nelson, K. H. Pepin, X. Qin, G. G. Sutton, E. Venter, B. P. Walenz, J. W. Wallis, K. C. Worley, S. P. Yang, S. M. Jones, M. A. Marra, M. Rocchi, J. E. Schein, R. Baertsch, L. Clarke, M. Csuros, J. Glasscock, R. A. Harris, P. Havlak, A. R. Jackson, H. Jiang, Y. Liu, D. N. Messina, Y. Shen, H. X. Song, T. Wylie, L. Zhang, E. Birney, K. Han, M. K. Konkel, J. Lee, A. F. Smit, B. Ullmer, H. Wang, J. Xing, R. Burhans, Z. Cheng, J. E. Karro, J. Ma, B. Raney, X. She, M. J. Cox, J. P. Demuth, L. J. Dumas, S. G. Han, J. Hopkins, A. Karimpour-Fard, Y. H. Kim, J. R. Pollack, T. Vinar, C. Addo-Quaye, J. Degenhardt, A. Denby, M. J. Hubisz, A. Indap, C. Kosiol, B. T. Lahn, H. A. Lawson, A. Marklein, R. Nielsen, E. J. Vallender, A. G. Clark, B. Ferguson, R. D. Hernandez, K. Hirani, H. Kehrer-Sawatzki, J. Kolb, S. Patil, L. L. Pu, Y. Ren, D. G. Smith, D. A. Wheeler, I. Schenck, E. V. Ball, R. Chen, D. N. Cooper, B. Giardine, F. Hsu, W. J. Kent, A. Lesk, D. L. Nelson, E. O'Brien W. K. Prufer, P. D. Stenson, J. C. Wallace, H. Ke, X. M. Liu, P. Wang, A. P. Xiang, F. Yang, G. P. Barber, D. Haussler, D. Karolchik, A. D. Kern, R. M. Kuhn, K. E. Smith, and A. S. Zwieg. 2007. Evolutionary and biomedical insights from the rhesus macaque genome. *Science* 316(5822):222-234.

GlaxoSmithKline. 2011. *Use of non-human primates (NHPs) in the discovery and development of medicines and vaccines.* http://www.gsk.com/policies/GSK-public-position-on-NHP.pdf (accessed July 27, 2011).

Goncalvez, A. P., R. Men, C. Wernly, R. H. Purcell, and C. J. Lai. 2004a. Chimpanzee fab fragments and a derived humanized immunoglobulin g1 antibody that efficiently cross-neutralize dengue type 1 and type 2 viruses. *J Virol* 78(23):12910-12918.

Goncalvez, A. P., R. H. Purcell, and C. J. Lai. 2004b. Epitope determinants of a chimpanzee fab antibody that efficiently cross-neutralizes dengue type 1 and type 2 viruses map to inside and in close proximity to fusion loop of the dengue type 2 virus envelope glycoprotein. *J Virol* 78(23):12919-12928.

Goncalvez, A. P., R. E. Engle, M. St. Claire, R. H. Purcell, and C. J. Lai. 2007. Monoclonal antibody-mediated enhancement of dengue virus infection in vitro and in vivo and strategies for prevention. *Proc Natl Acad Sci USA* 104(22):9422-9427.

Goncalvez, A. P., C. H. Chien, K. Tubthong, I. Gorshkova, C. Roll, O. Donau, P. Schuck, S. Yoksan, S. D. Wang, R. H. Purcell, and C. J. Lai. 2008. Humanized monoclonal antibodies derived from chimpanzee fabs protect against Japanese encephalitis virus in vitro and in vivo. *J Virol* 82(14):7009-7021.

Graham, B. S. 2011. Biological challenges and technological opportunities for respiratory syncytial virus vaccine development. *Immunol Rev* 239(1):149-166.

Halliday, J., P. Klenerman, and E. Barnes. 2011. Vaccination for hepatitis C virus: Closing in on an evasive target. *Expert Rev Vaccines* 10(5):659-672.

Harcourt, J. L., H. Caidi, L. J. Anderson, and L. M. Haynes. 2011. Evaluation of the calu-3 cell line as a model of in vitro respiratory syncytial virus infection. *J Virol Methods* 174(1-2):144-149.

Hartung, T. 2010. Comparative analysis of the revised directive 2010/63/EU for the protection of laboratory animals with its predecessor 86/609/EEC—a T4 report. *ALTEX* 27(4):285-303.

HHS (U.S. Department of Health and Human Services). 2005. *Code of Federal Regulations title 45 public welfare.* Washington, DC: U.S. National Institutes of Health, Office of Human Subjects Research.

HHS. 2011a. *Costs for maintaining humane care and welfare of chimpanzees.* http://grants.nih.gov/grants/policy/air/cost_for_caring_housing_of_chimpan zees_20110609.htm (accessed July 20, 2011).

HHS. 2011b. *NIH statement on the Alamogordo Primate Facility chimpanzees.* http://grants.nih.gov/grants/policy/air/nih_alamogordo_statement_20110104. htm (accessed September 9, 2011).

HHS. 2011c. *Operation and maintenance of a chimpanzee facility: NHLBI-csb-(rr)-ss-2011-264-kjm.* https://www.fbo.gov/spg/HHS/NIH/OAM/NHLBI-CSB-(RR)-SS-2011-264-KJM/listing.html (accessed August 24, 2011).

HHS. 2011d. *Response to the Physicians Committee for Responsible Medicine petition for administrative action April 8, 2011.* http://grants.nih.gov/grants/policy/air/physicians_committee_response_20110408.htm (accessed October 20, 2011).

Horner, V., J. D. Carter, M. Suchak, and F. B. de Waal. 2011. Spontaneous prosocial choice by chimpanzees. *Proc Natl Acad Sci USA* 108(33):13847-13851.

Hosea, N. A. 2011. Drug design tools—in silico, in vitro and in vivo ADME/PK prediction and interpretation: Is PK in monkey an essential part of a good human PK prediction? *Curr Top Med Chem* 11(4):351-357.

Houghton, M. 2011. Prospects for prophylactic and therapeutic vaccines against the hepatitis C viruses. *Immunol Rev* 239(1):99-108.

Houghton, M., and S. Abrignani. 2005. Prospects for a vaccine against the hepatitis C virus. *Nature* 436(7053):961-966.

The Humane Society of the United States. 2010. The HSUS urges Department of Health and Human Services to retire 202 chimpanzees. Press release, August 18. Gaithersburg, MD: The Humane Society of the United States.

ICH Harmonized Tripartite Guideline. 2011. *Preclinical safety evaluation of biotechnology derived pharmaceuticals S6(R1).* http://www.ich.org/fileadmin/ Public_Web_Site/ICH_Products/Guidelines/Safety/S6_R1/Step4/S6_R1_Guide line.pdf (accessed October 13, 2011).

Ilyas, J. A., and J. M. Vierling. 2011. An overview of emerging therapies for the treatment of chronic hepatitis C. *Clin Liver Dis* 15(3):515-536.

Ings, R. M. 2009. Microdosing: A valuable tool for accelerating drug development and the role of bioanalytical methods in meeting the challenge. *Bioanalysis* 1(7):1293-1305.

Iwarson, S., E. Tabor, H. C. Thomas, A. Goodall, J. Waters, P. Snoy, J. W. Shih, and R. J. Gerety. 1985. Neutralization of hepatitis B virus infectivity by a murine monoclonal antibody: An experimental study in the chimpanzee. *J Med Virol* 16(1):89-96.

Jacobson, I. M., J. G. McHutchison, G. Dusheiko, A. M. Di Bisceglie, K. R. Reddy, N. H. Bzowej, P. Marcellin, A. J. Muir, P. Ferenci, R. Flisiak, J. George, M. Rizzetto, D. Shouval, R. Sola, R. A. Terg, E. M. Yoshida, N. Adda, L. Bengtsson, A. J. Sankoh, T. L. Kieffer, S. George, R. S. Kauffman, and S. Zeuzem. 2011. Telaprevir for previously untreated chronic hepatitis C virus infection. *N Engl J Med* 364(25):2405-2416.

Jensen, D. M. 2011. A new era of hepatitis C therapy begins. *N Engl J Med* 364(13):1272-1274.

Kapoor, A., P. Simmonds, G. Gerold, N. Qaisar, K. Jain, J. A. Henriquez, C. Firth, D. L. Hirschberg, C. M. Rice, S. Shields, and W. I. Lipkin. 2011. Characterization of a canine homolog of hepatitis C virus. *Proc Natl Acad Sci USA* 108(28):11608-11613.

Kingstone, A., C. K. Friesen, and M. S. Gazzaniga. 2000. Reflexive joint attention depends on lateralized cortical connections. *Psychol Sci* 11(2):159-166.

Klevens, R. M., and R. A. Tohme. 2010. Evaluation of acute hepatitis C infection surveillance—United States, 2008. *MMWR* 59(43):4.

Kneteman, N. M., and D. F. Mercer. 2005. Mice with chimeric human livers: Who says supermodels have to be tall? *Hepatology* 41(4):703-706.

Köhler, G., and C. Milstein. 1975. Continuous cultures of fused cells secreting antibody of predefined specificity. *Nature* 256(5517):495-497.

Konopka, G., J. M. Bomar, K. Winden, G. Coppola, Z. O. Jonsson, F. Gao, S. Peng, T. M. Preuss, J. A. Wohlschlegel, and D. H. Geschwind. 2009. Human-specific transcriptional regulation of CNS development genes by FOXP2. *Nature* 462(7270):213-217.

Korte, T. 2010. Chimps' future prompts debate over NM primate lab. *Associated Press Online*. September 22.

Krilov, L. R. 2011. Respiratory syncytial virus disease: Update on treatment and prevention. *Expert Rev Anti Infect Ther* 9(1):27-32.

Kurosaki, M., N. Enomoto, F. Marumo, and C. Sato. 1993. Rapid sequence variation of the hypervariable region of hepatitis C virus during the course of chronic infection. *Hepatology* 18(6):1293-1299.

Kwo, P. Y., E. J. Lawitz, J. McCone, E. R. Schiff, J. M. Vierling, D. Pound, M. N. Davis, J. S. Galati, S. C. Gordon, N. Ravendhran, L. Rossaro, F. H. Anderson, I. M. Jacobson, R. Rubin, K. Koury, L. D. Pedicone, C. A. Brass, E. Chaudhri, and J. K. Albrecht. 2010. Efficacy of boceprevir, an NS3 protease inhibitor, in combination with peginterferon alfa-2b and ribavirin in treatment-naive patients with genotype 1 hepatitis C infection (sprint-1): An open-label, randomised, multicentre phase 2 trial. *Lancet* 376(9742):705-716.

Lai, C. S., S. E. Fisher, J. A. Hurst, F. Vargha-Khadem, and A. P. Monaco. 2001. A forkhead-domain gene is mutated in a severe speech and language disorder. *Nature* 413(6855):519-523.

Lammey, M. 2011. (Alamogordo Primate Facility, Alamogordo, NM). Email to authors: IOM chimp study with file attachment: APF colony age sex breakdown.xls. May 17, 2011.

Landry, J. 2011. (New Iberia Research Center, New Iberia, LA). Email to authors: IOM chimp study with file attachment: NIRC chimps count by age.xls. May 17, 2011.

Lanford, R. E., B. Guerra, C. B. Bigger, H. Lee, D. Chavez, and K. M. Brasky. 2007. Lack of response to exogenous interferon-alpha in the liver of chimpanzees chronically infected with hepatitis C virus. *Hepatology* 46(4):999-1008.

Lanford, R., B. Guerra, D. Chavez, V. L. Hodara, X. Zheng, G. Wolfgang, and D. B. Tumas. 2011. *Therapeutic efficacy of a TLR7 agonist for HBV chronic infection in chimpanzees*. Paper presented at 46th Annual Meeting of the European Association for the Study of the Liver, Berlin, Germany, March 30-April 3.

Langford, R. E. 2011. (Texas Biomedical Research Institute, San Antonio, TX). Email to authors: Chimpanzee data from SNPRC for IOM committee with file attachment: SNPRC chimpanzee studies and biomaterials IOM 62811.pdf. June 29, 2011.

Ledford, H. 2010. Chimps' fate ignites debate. *Nature* 467(7315):507-508.

Lindenbach, B. D., M. J. Evans, A. J. Syder, B. Wolk, T. L. Tellinghuisen, C. C. Liu, T. Maruyama, R. O. Hynes, D. R. Burton, J. A. McKeating, and C. M. Rice. 2005. Complete replication of hepatitis C virus in cell culture. *Science* 309(5734):623-626.

Lohmann, V., F. Korner, J. Koch, U. Herian, L. Theilmann, and R. Bartenschlager. 1999. Replication of subgenomic hepatitis C virus RNAs in a hepatoma cell line. *Science* 285(5424):110-113.

Ma, Y., L. Poisson, G. Sanchez-Schmitz, S. Pawar, C. Qu, G. J. Randolph, W. L. Warren, E. M. Mishkin, and R. G. Higbee. 2010. Assessing the immunopotency of toll-like receptor agonists in an in vitro tissue-engineered immunological model. *Immunology* 130(3):374-387.

MacDermot, K. D., E. Bonora, N. Sykes, A. M. Coupe, C. S. Lai, S. C. Vernes, F. Vargha-Khadem, F. McKenzie, R. L. Smith, A. P. Monaco, and S. E. Fisher. 2005. Identification of FOXP2 truncation as a novel cause of developmental speech and language deficits. *Am J Hum Genet* 76(6):1074-1080.

Maher, L., B. White, M. Hellard, A. Madden, M. Prins, T. Kerr, and K. Page. 2010. Candidate hepatitis C vaccine trials and people who inject drugs: Challenges and opportunities. *Vaccine* 28(45):7273-7278.

Marasco, W. A., and J. Sui. 2007. The growth and potential of human antiviral monoclonal antibody therapeutics. *Nat Biotechnol* 25(12):1421-1434.

Marques-Bonet, T., O. A. Ryder, and E. E. Eichler. 2009. Sequencing primate genomes: What have we learned? *Annu Rev Genomics Hum Genet* 10:355-386.

MedImmune LLC. 2011a. A randomized, double-blind, placebo-controlled study to evaluate viral shedding of medi-559 in healthy 1 to <24 month-old children. *ClinicalTrials.gov: A service of the U.S. National Institutes of Health.* http://clinicaltrials.gov/ct2/show/NCT00767416?term=NCT00767416&rank=1 NLM Idenitifier: NCT00767416 (accessed August 24, 2011).

MedImmune LLC. 2011b. A study to evaluate the safety, tolerability, immunogenicity and vaccine-like viral shedding of MEDI-534, against respiratory syncytial virus (RSV) and parainfluenza virus type 3 (PIV3), in healthy 6 to <24 month-old children and in 2 month-old infants. *ClinicalTrials.gov: A service of the U.S. National Institutes of Health.* http://clinicaltrials.gov/ct2/show/NCT00686075?term=NCT00686075&rank=1 NLM Identifier: NCT00686075 (accessed August 24, 2011).

Men, R., T. Yamashiro, A. P. Goncalvez, C. Wernly, D. J. Schofield, S. U. Emerson, R. H. Purcell, and C. J. Lai. 2004. Identification of chimpanzee fab fragments by repertoire cloning and production of a full-length humanized immunoglobulin g1 antibody that is highly efficient for neutralization of dengue type 4 virus. *J Virol* 78(9):4665-4674.

Mercer, D. F., D. E. Schiller, J. F. Elliott, D. N. Douglas, C. Hao, A. Rinfret, W. R. Addison, K. P. Fischer, T. A. Churchill, J. R. Lakey, D. L. Tyrrell, and N. M. Kneteman. 2001. Hepatitis C virus replication in mice with chimeric human livers. *Nat Med* 7(8):927-933.

Meuleman, P., L. Libbrecht, R. De Vos, B. de Hemptinne, K. Gevaert, J. Vandekerckhove, T. Roskams, and G. Leroux-Roels. 2005. Morphological and biochemical characterization of a human liver in a UPA-SCID mouse chimera. *Hepatology* 41(4):847-856.

Meunier, J. C., J. M. Gottwein, M. Houghton, R. S. Russell, S. U. Emerson, J. Bukh, and R. H. Purcell. 2011. Vaccine-induced cross-genotype reactive neutralizing antibodies against hepatitis C virus. *J Infect Dis* 204(8):1186-1190.

MicroDose Therapeutx. 2011. A trial to assess the safety, tolerability, and pharmacokinetics of MDT-637 in healthy volunteers. *ClinicalTrials.gov: A service of the U.S. National Institutes of Health.* http://clinicaltrials.gov/ct2/show/NCT01355016?term=nct01355016&rank=1. NLM Identifier: NCT01355016 (accessed August 24, 2011).

Mikkelsen, T. S., L. W. Hillier, E. E. Eichler, M. C. Zody, D. B. Jaffe, S.-P. Yang, W. Enard, I. Hellmann, K. Lindblad-Toh, T. K. Altheide, N. Archidiacono, P. Bork, J. Butler, J. L. Chang, Z. Cheng, A. T. Chinwalla, P. deJong, K. D. Delehaunty, C. C. Fronick, L. L. Fulton, Y. Gilad, G. Glusman, S. Gnerre, T. A. Graves, T. Hayakawa, K. E. Hayden, X. Huang, H. Ji, W. J. Kent, M.-C. King, E. J. KulbokasIII, M. K. Lee, G. Liu, C. Lopez-Otin, K. D. Makova, O. Man, E. R. Mardis, E. Mauceli, T. L. Miner, W. E. Nash, J. O. Nelson, S. Pääbo, N. J. Patterson, C. S. Pohl, K. S. Pollard, K. Prüfer, X. S. Puente, D. Reich, M. Rocchi, K. Rosenbloom, M. Ruvolo, D. J. Richter, S. F. Schaffner, A. F. A. Smit, S. M. Smith, M. Suyama, J. Taylor, D. Torrents, E. Tuzun, A. Varki, G. Velasco, M. Ventura, J. W. Wallis, M. C. Wendl, R. K. Wilson, E. S. Lander, and R. H. Waterston. 2005. Initial sequence of the chimpanzee genome and comparison with the human genome. *Nature* 437(7055):69-87.

Mohapatra, S. S., and S. Boyapalle. 2008. Epidemiologic, experimental, and clinical links between respiratory syncytial virus infection and asthma. *Clin Microbiol Rev* 21(3):495-504.

Muller, P. Y., M. Milton, P. Lloyd, J. Sims, and F. R. Brennan. 2009. The minimum anticipated biological effect level (MABEL) for selection of first human dose in clinical trials with monoclonal antibodies. *Curr Opin Biotechnol* 20(6):722-729.

Mundy, P., M. Gwaltney, and H. Henderson. 2010. Self-referenced processing, neurodevelopment and joint attention in autism. *Autism* 14(5):408-429.

Murphy, B. R., S. L. Hall, J. Crowe, P. L. Collins, E. K. Subbarao, M. Connors, W. T. London, and R. M. Chanock. 1992. The use of chimpanzees in respiratory virus research. In *Chimpanzee conservation and public health: Environments for the future,* 1st ed., edited by J. Erwin and J. C. Landon. Rockville, MD: Diagnon Corp./Bioqual. Pp. 21-27.

Nair, H., V. R. Verma, E. Theodoratou, L. Zgaga, T. Huda, E. A. Simoes, P. F. Wright, I. Rudan, and H. Campbell. 2011. An evaluation of the emerging interventions against respiratory syncytial virus (RSV)-associated acute lower respiratory infections in children. *BMC Public Health* 11(Suppl 3):S30.

NCRR (National Center for Research Resources). 2007. *Report of the chimpanzee management plan working group.* http://ncrr.nih.gov/comparative_medicine/

chimpanzee_management_program/ChimP05-22-2007.pdf (accessed July 27, 2011).

NCRR. 2011a. *Primate resources.* http://www.ncrr.nih.gov/comparative_medicine/resource_directory/primates.asp#alamo (accessed July 27, 2011).

NCRR. 2011b. *Chimpanzee management program.* http://www.ncrr.nih.gov/comparative_medicine/chimpanzee_mangementprogram/ (accessed July 27, 2011).

Nelson, A. L., E. Dhimolea, and J. M. Reichert. 2010. Development trends for human monoclonal antibody therapeutics. *Nat Rev Drug Discov* 9(10):767-774.

NIAID (U.S. National Institute of Allergy and Infectious Diseases). 2011. *Emerging and re-emerging infectious diseases.* http://www.niaid.nih.gov/topics/emerging/Pages/list.aspx (accessed August 18, 2011).

NIH. 2011. *About NIH: Mission.* http://nih.gov/about/mission.htm (accessed August 13, 2011).

NIH (U.S. National Institutes of Health). Unpublished. *Charter of the interagency animal model committee.*

Novavax. 2011. Safety study of respiratory syncytial virus F (RSV-F) vaccine to treat the respiratory syncytial virus in healthy adults 18 to 49 years of age (nvx 757 101). *ClinicalTrials.gov: A service of the U.S. National Institute of Health.* http://clinicaltrials.gov/ct2/show/NCT01290419?term=NCT01290419 NCT01290419&rank=1 NLM Identifier: NCT01290419 (accessed August 2, 2011).

NRC (National Research Council). 1997. *Chimpanzees in research: Strategies for their ethical care, management, and use.* Washington, DC: National Academy Press.

NRC. 2010. *Guide for the care and use of laboratory animals,* 8th ed. Washington, DC: The National Academies Press.

NSF (National Science Foundation). 2011. *Funding search results.* http://www.nsf.gov/funding/funding_results.jsp?nsfOrgs=allorg&pubStatus=ACTIVE&queryText=chimpanzee&searchAwards=on&Submit.x=0&Submit.y=0 (accessed July 14, 2011).

Okasha, S. 2010. Altruism researchers must cooperate. *Nature* 467(7316):653-655.

Olsen, D. B., M. E. Davies, L. Handt, K. Koeplinger, N. R. Zhang, S. W. Ludmerer, D. Graham, N. Liverton, M. MacCoss, D. Hazuda, and S. S. Carroll. 2011. Sustained viral response in a hepatitis C virus-infected chimpanzee via a combination of direct-acting antiviral agents. *Antimicrob Agents Chemother* 55(2):937-939.

Pan, C. H., C. E. Greer, D. Hauer, H. S. Legg, E. Y. Lee, M. J. Bergen, B. Lau, R. J. Adams, J. M. Polo, and D. E. Griffin. 2010. A chimeric alphavirus replicon particle vaccine expressing the hemagglutinin and fusion proteins protects juvenile and infant rhesus macaques from measles. *J Virol* 84(8):3798-3807.

Parliament of the United Kingdom. 1987. *Guidance on the Operation of the Animals (Scientific Procedures) Act 1986.* On January 1, 1997. Act Eliz. II 1986 C.14 Section 21. United Kingdom.

Pharmaceutical Business Review. 2011. *Search of clinical trials for preclinical RSV programs in development for either therapeutics or prophylactics.* http://clinicaltrials.pharmaceutical-business-review.com (accessed August 24, 2011).

Physicians Committee for Responsible Medicine. 2011. Petition for administrative action before the National Institutes of Health, submitted to: Francis S. Collins, Director, National Institutes of Health, March 3.

Ploss, A., and C. M. Rice. 2009. Towards a small animal model for hepatitis C. *EMBO Rep* 10(11):1220-1227.

Pollack, P., and J. R. Groothuis. 2002. Development and use of palivizumab (synagis): A passive immunoprophylactic agent for RSV. *J Infect Chemother* 8(3):201-206.

Prescott, W. A., Jr., F. Doloresco, J. Brown, and J. A. Paladino. 2010. Cost effectiveness of respiratory syncytial virus prophylaxis: A critical and systematic review. *Pharmacoeconomics* 28(4):279-293.

Qiu, J., B. Qin, S. Rayner, C. C. Wu, R. J. Pei, S. Xu, Y. Wang, and X. W. Chen. 2011. Novel evidence suggests hepatitis B virus surface proteins participate in regulation of HBV genome replication. *Virol Sin* 26(2):131-138.

Reichert, J. M., C. J. Rosensweig, L. B. Faden, and M. C. Dewitz. 2005. Monoclonal antibody successes in the clinic. *Nat Biotechnol* 23(9):1073-1078.

Reynolds, J., and CEECE (Coalition to End Experiments on Chimpanzees in Europe). 2001. *Report on the Biomedical Primate Research Centre (BPRC).* Brighton, UK: CEECE.

Reynolds, T. 2011. *Monoclonal antibody therapeutics* [slide presentation]. Presented to the Committee on the Use of Chimpanzees in Biomedical and Behavioral Research, August 11, 2011, Washington, DC.

Rowell, T. J. 2011. (New Iberia Research Center, New Iberia, LA). Email to authors: Chimpanzee data from NIRC for IOM committee with file attachment: studies and bio requests.pdf. June 30, 2011.

Sakai, T., A. Mikami, M. Tomonaga, M. Matsui, J. Suzuki, Y. Hamada, M. Tanaka, T. Miyabe-Nishiwaki, H. Makishima, M. Nakatsukasa, and T. Matsuzawa. 2011. Differential prefrontal white matter development in chimpanzees and humans. *Curr Biol* 21(16):1397-1402.

SCHER (Scientific Committee on Health and Environmental Risks). 2009. *The need for non-human primates in biomedical research, production and testing of products and devices.* Adopted at the 27th plenary on 13 January 2009. European Commission, Health and Protection Consumer Directorate-General. Brussels.

Schofield, D. J., J. Glamann, S. U. Emerson, and R. H. Purcell. 2000. Identification by phage display and characterization of two neutralizing chimpanzee monoclonal antibodies to the hepatitis E virus capsid protein. *J Virol* 74(12):5548-5555.

Schofield, D. J., W. Satterfield, S. U. Emerson, and R. H. Purcell. 2002. Four chimpanzee monoclonal antibodies isolated by phage display neutralize hepatitis A virus. *Virology* 292(1):127-136.

Schofield, D. J., R. H. Purcell, H. T. Nguyen, and S. U. Emerson. 2003. Monoclonal antibodies that neutralize HEVrecognize an antigenic site at the carboxyterminus of an ORF2 protein vaccine. *Vaccine* 22(2):257-267.

Secretary of State for the Home Department and Parliament of the United Kingdom. 1998. *Report of the Animal Procedures Committee for 1997.* London, UK: Secretary of State for the Home Department and Parliament.

Shadman, K. A., and E. R. Wald. 2011. A review of palivizumab and emerging therapies for respiratory syncytial virus. *Expert Opin Biol Ther*, http://www.ncbi.nlm.nih.gov/pubmed/21831008 (accessed August 11, 2011).

Shah, J. N., and R. F. Chemaly. 2011. Management of RSV infections in adult recipients of hematopoietic stem cell transplantation. *Blood* 117(10):2755-2763.

Sheehy, P., B. Mullan, I. Moreau, E. Kenny-Walsh, F. Shanahan, M. Scallan, and L. J. Fanning. 2007. In vitro replication models for the hepatitis C virus. *J Viral Hepat* 14(1):2-10.

Shoukry, N. H., A. Grakoui, M. Houghton, D. Y. Chien, J. Ghrayeb, K. A. Reimann, and C. M. Walker. 2003. Memory CD8+ T cells are required for protection from persistent hepatitis C virus infection. *J Exp Med* 197(12):1645-1655.

Strickland, G. T., S. S. El-Kamary, P. Klenerman, and A. Nicosia. 2008. Hepatitis C vaccine: Supply and demand. *Lancet Infect Dis* 8(6):379-386.

Su, A. I., J. P. Pezacki, L. Wodicka, A. D. Brideau, L. Supekova, R. Thimme, S. Wieland, J. Bukh, R. H. Purcell, P. G. Schultz, and F. V. Chisari. 2002. Genomic analysis of the host response to hepatitis C virus infection. *Proc Natl Acad Sci USA* 99(24):15669-15674.

Suomi, S. J. 2006. Risk, resilience, and gene X environment interactions in rhesus monkeys. *Ann N Y Acad Sci* 1094:52-62.

Tabrizi, M. A., G. G. Bornstein, S. L. Klakamp, A. Drake, R. Knight, and L. Roskos. 2009. Translational strategies for development of monoclonal antibodies from discovery to the clinic. *Drug Discov Today* 14(5-6):298-305.

Taglialatela, J. P., J. L. Russell, J. A. Schaeffer, and W. D. Hopkins. 2008. Communicative signaling activates "Broca's" homolog in chimpanzees. *Curr Biol* 18(5):343-348.

Tayyari, F., D. Marchant, T. J. Moraes, W. Duan, P. Mastrangelo, and R. G. Hegele. 2011. Identification of nucleolin as a cellular receptor for human respiratory syncytial virus. *Nat Med* 17(9):1132-1135.

Thimme, R., J. Bukh, H. C. Spangenberg, S. Wieland, J. Pemberton, C. Steiger, S. Govindarajan, R. H. Purcell, and F. V. Chisari. 2002. Viral and immunological determinants of hepatitis C virus clearance, persistence, and disease. *Proc Natl Acad Sci USA* 99(24):15661-15668.

Thomson, M., M. Nascimbeni, M. B. Havert, M. Major, S. Gonzales, H. Alter, S. M. Feinstone, K. K. Murthy, B. Rehermann, and T. J. Liang. 2003. The clearance of hepatitis C virus infection in chimpanzees may not necessarily correlate with the appearance of acquired immunity. *J Virol* 77(2):862-870.

U.S. Office of Laboratory Animal Welfare. 2002. *Public Health Service policy on humane care and use of laboratory animals.* http://grants.nih.gov/grants/ http://grants.nih.gov/grants/olaw/references/phspol.htm (accessed August 18, 2011).

Valenzuela, D. M., A. J. Murphy, D. Frendewey, N. W. Gale, A. N. Economides, W. Auerbach, W. T. Poueymirou, N. C. Adams, J. Rojas, J. Yasenchak, R. Chernomorsky, M. Boucher, A. L. Elsasser, L. Esau, J. Zheng, J. A. Griffiths, X. Wang, H. Su, Y. Xue, M. G. Dominguez, I. Noguera, R. Torres, L. E. Macdonald, A. F. Stewart, T. M. DeChiara, and G. D. Yancopoulos. 2003. High-throughput engineering of the mouse genome coupled with high-resolution expression analysis. *Nat Biotechnol* 21(6):652-659.

Valerio, L. G., Jr. 2011. In silico toxicology models and databases as FDA Critical Path Initiative toolkits. *Hum Genomics* 5(3):200-207.

Verstrepen, B. E., E. Depla, C. S. Rollier, G. Mares, J. A. Drexhage, S. Priem, E. J. Verschoor, G. Koopman, C. Granier, M. Dreux, F. L. Cosset, G. Maertens, and J. L. Heeney. 2011. Clearance of genotype 1B hepatitis C virus in chimpanzees in the presence of vaccine-induced E1-neutralizing antibodies. *J Infect Dis* 204(6):837-844.

Vertex Pharmaceuticals Incorporated. 2011. *Telaprevir 375-mg film-coated tablet for the treatment of genotype 1 chronic hepatitis C.* hhtp://www.fda.gov/down loads/AdvisoryCommittees/CommitteesMeetingMaterials/Drugs/AntiviralDrugs AdvisoryCommittee/UCM252562.pdf (accessed September 13, 2011).

Vogel, G. 2001. Dutch end chimp studies. *ScienceNOW.* http://news.sciencemag. http://news.sciencemag.org/sciencenow/2001/05/01-01.html (accessed September 30, 2011).

Wang, C. C., E. Krantz, J. Klarquist, M. Krows, L. McBride, E. P. Scott, T. Shaw-Stiffel, S. J. Weston, H. Thiede, A. Wald, and H. R. Rosen. 2007. Acute hepatitis C in a contemporary U.S. cohort: Modes of acquisition and factors influencing viral clearance. *J Infect Dis* 196(10):1474-1482.

Waterston, R. H., K. Lindblad-Toh, E. Birney, J. Rogers, J. F. Abril, P. Agarwal, R. Agarwala, R. Ainscough, M. Alexandersson, P. An, S. E. Antonarakis, J. Attwood, R. Baertsch, J. Bailey, K. Barlow, S. Beck, E. Berry, B. Birren, T. Bloom, P. Bork, M. Botcherby, N. Bray, M. R. Brent, D. G. Brown, S. D. Brown, C. Bult, J. Burton, J. Butler, R. D. Campbell, P. Carninci, S. Cawley, F. Chiaromonte, A. T. Chinwalla, D. M. Church, M. Clamp, C. Clee, F. S. Collins, L. L. Cook, R. R. Copley, A. Coulson, O. Couronne, J. Cuff, V. Curwen, T. Cutts, M. Daly, R. David, J. Davies, K. D. Delehaunty, J. Deri, E. T. Dermitzakis, C. Dewey, N. J. Dickens, M. Diekhans, S. Dodge,

I. Dubchak, D. M. Dunn, S. R. Eddy, L. Elnitski, R. D. Emes, P. Eswara, E. Eyras, A. Felsenfeld, G. A. Fewell, P. Flicek, K. Foley, W. N. Frankel, L. A. Fulton, R. S. Fulton, T. S. Furey, D. Gage, R. A. Gibbs, G. Glusman, S. Gnerre, N. Goldman, L. Goodstadt, D. Grafham, T. A. Graves, E. D. Green, S. Gregory, R. Guigo, M. Guyer, R. C. Hardison, D. Haussler, Y. Hayashizaki, L. W. Hillier, A. Hinrichs, W. Hlavina, T. Holzer, F. Hsu, A. Hua, T. Hubbard, A. Hunt, I. Jackson, D. B. Jaffe, L. S. Johnson, M. Jones, T. A. Jones, A. Joy, M. Kamal, E. K. Karlsson, D. Karolchik, A. Kasprzyk, J. Kawai, E. Keibler, C. Kells, W. J. Kent, A. Kirby, D. L. Kolbe, I. Korf, R. S. Kucherlapati, E. J. Kulbokas, D. Kulp, T. Landers, J. P. Leger, S. Leonard, I. Letunic, R. Levine, J. Li, M. Li, C. Lloyd, S. Lucas, B. Ma, D. R. Maglott, E. R. Mardis, L. Matthews, E. Mauceli, J. H. Mayer, M. McCarthy, W. R. McCombie, S. McLaren, K. McLay, J. D. McPherson, J. Meldrim, B. Meredith, J. P. Mesirov, W. Miller, T. L. Miner, E. Mongin, K. T. Montgomery, M. Morgan, R. Mott, J. C. Mullikin, D. M. Muzny, W. E. Nash, J. O. Nelson, M. N. Nhan, R. Nicol, Z. Ning, C. Nusbaum, M. J. O'Connor, Y. Okazaki, K. Oliver, E. Overton-Larty, L. Pachter, G. Parra, K. H. Pepin, J. Peterson, P. Pevzner, R. Plumb, C. S. Pohl, A. Poliakov, T. C. Ponce, C. P. Ponting, S. Potter, M. Quail, A. Reymond, B. A. Roe, K. M. Roskin, E. M. Rubin, A. G. Rust, R. Santos, V. Sapojnikov, B. Schultz, J. Schultz, M. S. Schwartz, S. Schwartz, C. Scott, S. Seaman, S. Searle, T. Sharpe, A. Sheridan, R. Shownkeen, S. Sims, J. B. Singer, G. Slater, A. Smit, D. R. Smith, B. Spencer, A. Stabenau, N. Stange-Thomann, C. Sugnet, M. Suyama, G. Tesler, J. Thompson, D. Torrents, E. Trevaskis, J. Tromp, C. Ucla, A. Ureta-Vidal, J. P. Vinson, A. C. Von Niederhausern, C. M. Wade, M. Wall, R. J. Weber, R. B. Weiss, M. C. Wendl, A. P. West, K. Wetterstrand, R. Wheeler, S. Whelan, J. Wierzbowski, D. Willey, S. Williams, R. K. Wilson, E. Winter, K. C. Worley, D. Wyman, S. Yang, S. P. Yang, E. M. Zdobnov, M. C. Zody, and E. S. Lander. 2002. Initial sequencing and comparative analysis of the mouse genome. *Nature* 420(6915):520-562.

Watson, H. 2011. *Chimpanzees in biomedical research* [slide presentation]. Presented to the Committee on the Use of Chimpanzees in Biomedical and Behavioral Research, May 26, 2011, Washington, DC.

Weisman, L. E. 2009. Respiratory syncytial virus (RSV) prevention and treatment: Past, present, and future. *Cardiovasc Hematol Agents Med Chem* 7(3):223-233.

WHO (World Health Organization). 2003. *Netherlands. Law of 2 October 2003 (stb. 399) amending the law on animal experimentation.* International Digest of Health Legislation. http://apps.who.int/idhl-rils/results.cfm?language=english&type=ByIssue&intDigestVolume=54&intIssue=4&strTopicCode=XIA (accessed August 24, 2011).

WHO. 2011. *Hepatitis C.* World Health Organization (WHO) Media Centre. Geneva.

Williams, I. T., B. P. Bell, W. Kuhnert, and M. J. Alter. 2011. Incidence and transmission patterns of acute hepatitis C in the United States, 1982-2006. *Arch Intern Med* 171(3):242-248.

Wright, M., and G. Piedimonte. 2011. Respiratory syncytial virus prevention and therapy: Past, present, and future. *Pediatr Pulmonol* 46(4):324-347.

Wu, G. Y., M. Konishi, C. M. Walton, D. Olive, K. Hayashi, and C. H. Wu. 2005. A novel immunocompetent rat model of HCV infection and hepatitis. *Gastroenterology* 128(5):1416-1423.

Yerkes National Primate Research Center. 2011. *About.* http://www.yerkes.emory.edu/about/index.html (accessed July 27, 2011).

Zanetti, A. R., E. Tanzi, S. Paccagnini, N. Principi, G. Pizzocolo, M. L. Caccamo, E. D'Amico, G. Cambie, and L. Vecchi. 1995. Mother-to-infant transmission of hepatitis C virus. Lombardy study group on vertical HCV transmission. *Lancet* 345(8945):289-291.

Zhong, J., P. Gastaminza, G. Cheng, S. Kapadia, T. Kato, D. R. Burton, S. F. Wieland, S. L. Uprichard, T. Wakita, and F. V. Chisari. 2005. Robust hepatitis C virus infection in vitro. *Proc Natl Acad Sci USA* 102(26):9294-9299.

B

Commissioned Paper: Comparison of Immunity to Pathogens in Humans, Chimpanzees, and Macaques

The following paper was commissioned by the Committee on the Use of Chimpanzees in Biomedical and Behavioral Research. The responsibility for the content of this paper rests with the authors and does not necessarily represent the views of the Institute of Medicine or its committees and convening bodies.

By: Nancy L. Haigwood, Ph.D.
Professor of Microbiology and Immunology
Director
Oregon National Primate Research Center

Christopher M. Walker, Ph.D.
Professor of Pediatrics
Nationwide Children's Hospital
The Ohio State University

INTRODUCTION

The purpose of this white paper is to compare genetic and functional features of immunity and the response to infection in humans and major nonhuman primate species currently used in biomedical research. The search for appropriate disease models has been stimulated by the need to understand the most intractable of the persistent and lethal pathogens, as well as chronic diseases and conditions that are determined by the genetic makeup of the individual. Because the outcome of infection is governed by carefully coordinated innate and adaptive immune responses, and pathogens have evolved strategies to evade these defenses, use of animal models that recapitulate key features of human infection is critical. Successful nonhuman primate models closely emulate human immunity, inflammation, and disease sequelae. They can also provide a critical pathway to clinical testing of risky prevention or treatment strategies for serious human diseases.

Some past successes of infectious diseases research in nonhuman primates are described. However, the primary objective of the paper is to identify conditions that either support or limit use of these animals for the study of human viral, bacterial, or parasitic infections. A survey of the published literature reveals that the common chimpanzee (*Pan troglodytes*) is the only great ape used in infectious disease research. With few exceptions there is usually no alternative, because lower species are not permissive for infection or fail to replicate key features of disease. Most studies involve very small numbers of chimpanzees to ensure safe translation of vaccines or therapeutics to humans, or provide incontrovertible evidence for basic mechanisms of immune control and evasion that cannot be obtained in human subjects. Various monkey species, primarily the Asian macaques (*Macaca species*), have served as models for infection with human viruses and microbes. Alternatively, monkey pathogens like the simian immunodeficiency viruses (SIV) provide a reliable model of human infection with closely related viruses like human immunodeficiency virus (HIV). Infection studies with human and monkey viruses have facilitated advances in vaccine development and studies of immunity and pathogenesis relevant to humans.

Sequencing of the human, chimpanzee, and macaque genomes has provided unprecedented insight into the evolutionary relationship between these species, especially for genes that regulate host defense and susceptibility to infection. Here we also provide examples of gene families involved in immunity that have been largely conserved since speciation, and others that have undergone rapid evolution because of selective pressure by infectious diseases. How these adaptive changes might affect modeling of human infectious diseases in monkeys and great apes is discussed. Contemporary examples of primate infectious disease models that replicate most if not all features of human infection and immunity are provided. Where infection models are not perfectly matched in humans and nonhuman primates, differences have provided insight into key features of the relationship between the pathogen and its human host.

Finally, several practical advantages of nonhuman primate models are also reviewed. The include the ability to (1) infect with clonal or genetically modified pathogens, (2) modify the immune response to identify protective mechanisms, (3) sample at the earliest times after infection, often before symptoms are apparent in humans, and (4) access organs or tissues that are the primary site of infection. The latter is particularly important because blood, which is often the only compartment available for human sampling, may not adequately reflect immunity at the site of infection. Advances in adapting the most sophisticated technologies to nonhuman primates, including methods to monitor immunity, and understand molecular aspects of infection using genomic and proteomic approaches, has the potential to provide new insight into vaccination and infection with known and emerging pathogens.

CHIMPANZEES

Historical and Current Examples of Human Infectious Disease Research in Chimpanzees

Chimpanzees have been used for over 100 years to model human viral, bacterial, and parasitic infections. This long history has revealed that chimpanzees are often uniquely permissive for infection with some medically important human pathogens. These animals can also provide a more faithful model of human disease than lower nonhuman primates. Studies in chimpanzees, particularly with hepatotropic viruses, have provided critical insight into host defense mechanisms and facilitated development of vaccines that have changed global public health. Yet for other pathogens key features of immunity and infection outcome differ between humans and chimpanzees. In these instances, the host-pathogen interaction is influenced by adaptations that are species-specific despite a strikingly close genetic relationship.

The promise and limitations of chimpanzees as an infectious disease model were first recognized in a 1904 publication from Albert Grunbaum who transmitted the Eberth-Gaffky bacillus (Salmonella typhi) to two animals (Grunbaum, 1904). The bacillus, isolated 10 years earlier, was the suspected cause of enteric fever. Efforts to satisfy Koch's third postulate by transmission of disease to rats, rabbits, and monkeys had failed. Infection of chimpanzees was successful, but with much milder disease symptoms than expected. The author noted "the virulence of my cultures was not sufficient to produce a fatal result in the two instances in which they were given the opportunity to do so" (Grunbaum, 1904). That the chimpanzee might not be suitable for S. typhi vaccine development was highlighted in a 1914 publication (Nichols, 1914). The author, a proponent of a killed vaccine, critiqued an earlier study where such an approach had failed. "The authors found that a whole killed vaccine did not protect chimpanzees. But they used tremendous infecting doses—the contents of a whole Kolle flask. The problems must be settled, as some of them already have been settled, by actual experience with large numbers of men kept under close observation" (Nichols, 1914).

Infectious disease research involving chimpanzees published in the last 30-40 years fits into three broad categories. They include (1) identification and characterization of infectious agents that are serious public health threats; (2) characterization of protective immunity and how it is subverted; and (3) development of strategies to prevent or treat human infections. All published experimental infection studies used human pathogens and not closely related (and thus potentially different) chimpanzee pathogens as a model. Here, factors that determine the suitability of chimpanzees for research on infectious diseases are reviewed. Malaria, respiratory syncytial virus (RSV), HIV, and the hepatitis viruses are used as case studies.

Malaria

Malaria vaccine research is made difficult by complexities of the parasite life cycle and selection of antigens to either interrupt transmission of infection or protect from disease after a mosquito bite (Good and Doolan, 2010). That irradiated *P. falciparum* sporozoites prevent disease was established several decades ago, but this approach is not easily scaled for human vaccination (Hoffman and Doolan, 2000). A small pilot study demonstrating protection of chimpanzees by a recombinant liver stage antigen derived from the sporozoites (Daubersies et al., 2000,

2008) laid the foundation for a recent human clinical trial of this concept (ClinicalTrials.gov NCT00509158). Nevertheless, malaria is exceptional because human challenge studies are permissible and studies in lower-order species can provide guidance for vaccine development. Many malaria trials that are either underway or completed involved parasite challenge of vaccinated human volunteers followed by co-artemether eradication therapy if necessary (as an example see Porter et al., 2011). Recent studies have also included sophisticated analyses of humoral and cellular immune responses with goal of identifying protective correlates in human volunteers (Good, 2011). While chimpanzees have been used sparingly to date, there is increasing concern that no successful vaccine has emerged from the concepts tested to date. If progress requires identification of new antigens and a better understanding of immunity, especially in the liver (Good, 2011), the place of the chimpanzee in malaria research may be reconsidered.

Chimpanzees have also been critically important for the in vivo generation of malaria parasites that are recombinants between drug-resistant and drug-sensitive strains in order to map drug resistance genes and thereby better understand the metabolism of these pathogens and to develop improved drugs. Several studies that used parasites generated by this approach have been published recently (Hayton et al., 2008; Nguitragool et al., 2011; Sa et al., 2009).

Respiratory Syncytial Virus

RSV was first isolated from captive chimpanzees with upper respiratory tract disease (Blount et al., 1956) but was quickly identified as a human virus (Chanock and Finberg, 1957; Chanock et al., 1957). It is now recognized as the most important viral agent of severe respiratory tract disease in infants and children worldwide (Hall et al., 2009; Nair et al., 2010). RSV is also an important cause of morbidity and mortality in the elderly and in profoundly immunosuppressed individuals. Protection of vulnerable infants and young children from severe airway disease by a licensed monoclonal antibody against the RSV F protein, as well as the protection in the general population afforded by prior infection, suggests that active vaccination is also feasible (Graham, 2011). Progress over the past 4 decades was slowed by an unfortunate clinical trial of a formalin inactivated RSV vaccine that worsened disease and resulted in two deaths upon natural infection with the virus (Kapikian et al., 1969). There is a general consensus that the vaccine failed to induce potent neutralizing antibody responses, provoked heightened lymphoproliferative responses, and was associated with eosinophilia and immune complex deposition in airways. Rodent and monkey models have demonstrated Th2 responses and eosinophilia using formalin-inactivated vaccines, but the precise mechanisms of immunopathogenesis remain undefined (reviewed in Graham, 2011). With the failure of this killed vaccine, development of live-attenuated RSV vaccines was initiated.

Chimpanzees are the only experimental animal in which RSV replication and pathogenicity approach that of humans. Small numbers of chimpanzees were used to demonstrate the safety of live-attenuated vaccines as well as identifying candidates that were sufficiently attenuated to move forward into clinical trials (e.g., Clinicaltrials.gov NCT00767416) (Crowe et al., 1993, 1994; Teng et al., 2000; Whitehead et al., 1999). Importantly, the body temperature of the chimpanzee is the same as that of humans. Other available nonhuman primates have higher body temperatures and so chimpanzees are uniquely suited for pre-clinical evaluation of temperature-sensitive vaccine candidates, which comprise all of the candidates evaluated in clinical trials to date. In addition, the chimpanzee experiments added to a body of evidence that both live attenuated vaccines and vectored vaccines are not associated with enhanced disease. A series of clinical trials have been initiated in infants and young children to evaluate safety, attenuation, and immunogenicity of several live RSV vaccines (for instance, see ClinicalTrials.gov NCT00767416). It is too soon to know if live RSV vaccines that are sufficiently attenuated to be well tolerated in young infants (Karron et al., 2005) will be sufficiently immunogenic to prevent severe RSV disease. Alternate approaches involving recombinant viral vectors (such as attenuated parainfluenza virus type 3; see ClinicalTrials.gov NCT00686075), virus-like particles (ClinicalTrials.gov NCT01290419), and subunit proteins are at various stages of development and evaluation. RSV subunit vaccines are considered unsuitable for use in RSV-naïve individuals, based on studies in mice, cotton rats, and African green monkeys. However, the evolutionary distance relative to humans, reduced permissiveness to RSV replication, and lack of disease may limit the predictive value of these models (Graham, 2011). Until the significant medical need for an RSV vaccine is fully met, it is difficult to exclude the possibility that chimpanzees will be required to answer questions about mechanisms and duration of immune protection and disease potentiation.

The Human Immunodeficiency Virus

Susceptibility of chimpanzees to persistent HIV infection was first reported in 1984, a little more than 1 year after discovery of the virus (Alter et al., 1984). Successful infection of two animals, and persistence of lymphadenopathy for several weeks in one of them, suggested that the chimpanzee would be valuable for further studies of acquired immune deficiency syndrome (AIDS) (Alter et al., 1985). This study, and all other early studies of immunity and vaccine development, used viruses like HIV_{IIIb} and HIV_{SF2} that adapted in culture to use CXCR4 as a co-receptor for cell entry. These viruses did initiate infection in chimpanzees, but viremia was usually low or short-lived and immunodeficiency was not observed (with one notable exception described below). Several studies of the early studies with CXCR4 adapted viruses nonetheless provided insight into the nature of antiviral immunity in infected chimpanzees (Nara et al., 1987), including the development of neutralizing antibodies (Prince et al., 1987), proliferative responses, and lack of $CD4^+$ T cell impairment (Eichberg et al., 1987). Important advances were made in understanding mucosal routes of HIV-1 transmission using chimpanzees (Fultz et al., 1986), as well as the now better-appreciated issue of superinfection (Fultz et al., 1987). Some success in protecting animals from infection with the CXCR4-dependent HIV strains was achieved by active and passive vaccination. Sterilizing immunity was induced by immunization with recombinant subunit envelope glycoproteins, but only with the CXCR4-utilizing virus HIV_{IIIB} matched to the envelope immunogen (Berman et al., 1988). The vaccine was not fully protective when a different strain belonging to the same subtype, HIV_{SF2}, was used (El-Amad et al., 1995). Passive transfer of HIVIG at a higher dose eventually showed protection against HIV_{IIIB} (Prince et al., 1991), as did one of the first neutralizing monoclonal antibodies directed against the HIV envelope (Emini et al., 1990). The interpretation of vaccine and antibody-based protection work in chimpanzees was complicated by the observation that chimpanzees were mostly resistant to infection with primary human HIV isolate that required the CCR5 chemokine receptor for cell entry. This discovery brought pause to the vaccine field, since none of the vaccines in testing could elicit neutralizing antibodies against the primary isolates that required CCR5 for infection. Later attempts to block infection using a human monoclonal against a primary HIV-1 challenge were also less successful, though they showed some effect in reducing acute phase viremia (Conley et al., 1996).

Notably, one chimpanzee did develop immunodeficiency after more than a decade of subclinical infection with two prototype strains of CXCR4-adapted HIV (Novembre et al., 1997). Isolation of a pathogenic virus from this animal sparked a debate on the role of chimpanzees in HIV vaccine research. Specifically, it was proposed that a pathogenic virus could facilitate a direct test of vaccines designed to slow $CD4^+$ T cell loss and immunodeficiency, if not infection (Cohen, 1999; Letvin, 1998). Vaccinated chimpanzees were never challenged with this virus in the face of persuasive ethical and scientific arguments (Cohen, 1999; Prince and Andrus, 1998). From the scientific perspective, there was considerable doubt about whether very rapid $CD4^+$ T cell loss observed after transmission of the virus to new animals was representative of human disease. Moreover, because immunodeficiency was not a consistent finding, there were also practical concerns with design of a study involving two or three animals per group. These experimental infections with the pathogenic HIV strain were perhaps the last conducted at a primate research facility in the United States. No new HIV infection studies in chimpanzees have been published in the past decade.

Studies on the origin of human HIV infection are beginning to yield insight into a host-virus relationship so finely tuned that it cannot be recapitulated in an animal with 99 percent genetic identity. It is now apparent that HIV originated from a chimpanzee simian immunodeficiency virus (SIVcpz) introduced into human populations by zoonotic infection at least three times since the beginning of the 20th century (Keele et al., 2009a). Most simian retroviruses, including those from monkeys, are restricted from growth in human cells by species-specific factors (see section below on Monkey-human models of infection). SIVcpz is no exception, as Vpu and Nef proteins had to adapt to neutralize human tetherin, a protein that is induced by interferon and restricts virus release from infected cells (Lim et al., 2010; Sauter et al., 2009). Adaptations like this one may explain attenuation of HIV infectivity for chimpanzees and perhaps limit its value as a model for vaccine development.

The Hepatitis Viruses

Chimpanzees are currently used to study the host response to four hepatitis viruses (HAV, HBV, HCV, and HEV) and to develop or refine approaches for prevention and treatment of the liver disease that they cause. Prevention and treatment of transmissible hepatitis in humans has been a public health priority for over 60 years. Progress toward isolation of the agent(s) responsible for the disease burden was slow, in part because hepatotropic viruses are fastidious and not easily propogated in cell culture. By the 1960s there was strong clinical, epidemiological, and immunological evidence for two distinct forms of transmissible hepatitis in humans (Krugman and Giles, 1972). Type A (or infectious) hepatitis had a short incubation period and was self-limited. Type B (or serum) hepatitis had a longer incubation period and was characterized by the prolonged presence of the Australia antigen (hepatitis B surface antigen; HBsAg) in serum. Use of chimpanzees in hepatitis research predated the discovery of the viruses that caused type A and B hepatitis. Sporadic outbreaks of liver disease in chimpanzee colonies with occasional zoonotic transmission indicated that the animals might be susceptible to infection with human viruses (Maynard et al., 1972a). Transmission of type A (Dienstag et al., 1975; Maynard et al., 1975) and B (Barker et al., 1973; Maynard et al., 1972b) hepatitis to chimpanzees facilitated characterization of both viruses and rapid development of diagnostics and highly effective vaccines. Chimpanzee research provided critical proof that HBV infection was preventable by vaccination with HBsAg purified from the serum of human carriers (Buynak et al., 1976a, 1976b; Purcell and Gerin, 1975), and for the transition to a safer recombinant subunit vaccine (McAleer et al., 1984). Attenuated and inactivated vaccines were also shown to prevent HAV infection of chimpanzees (Feinstone et al., 1983; Provost et al., 1983; Purcell et al., 1992). The principle that a vaccine can prevent disease when administered as post-exposure prophylaxis was also established using HAV-infected chimpanzees (Purcell et al., 1992). Universal childhood vaccination against HAV and HBV is now recommended in the United States.

Chimpanzees were also critically important to the discovery of the agent causing a third major form of human hepatitis. Studies published in 1975 concluded that unidentified type C hepatitis virus(es) were responsible for post-transfusion hepatitis in subjects not infected with HAV or HBV (Feinstone et al., 1975; Prince et al., 1974). The infectious nature of non-A, non-B hepatitis was not established by experimental transmission of liver disease to uninfected humans, an approach used to define features of type A and B hepatitis in the highly controversial Willowbrook experiments (Krugman, 1986). Instead, evidence that the disease was caused by a small, enveloped RNA virus was obtained by physico-chemical analysis of patient serum that transmitted persistent hepatitis to chimpanzees (Alter et al., 1978; Bradley et al., 1983, 1985; Tabor et al., 1978). Serum that was titrated and serially passaged in chimpanzees provided the pedigreed material from which the HCV genome was eventually cloned as described in 1989 (Choo et al., 1989).

All four hepatitis viruses remain significant public health problems today. A box summarizing research objectives of contemporary hepatitis virus research using chimpanzees is provided (Box 1). As described in detail below, HAV, HBV, HCV, and HEV all cause robust infection in chimpanzees. These viruses can cause the same spectrum of liver disease observed in humans, although in both species most infections tend to clinically mild and/or slowly progressive.

BOX 1
Major Uses of Chimpanzees in Infectious Disease Research

- Characterize and identify new infectious agents, especially those that cannot be propagated in lower species or cell culture.

- Define mechanisms of protective innate and adaptive immunity and pathogen evasion strategies. This is particularly important in settings where early phases of acute infection are not easily identified in humans, or infected tissues are not accessibly for studies of immunity.

- Establish that new concepts for vaccination or therapy of infection are safe and effective before translation to humans.

- Determine if reagents critical to development of therapeutics like clonal viruses or parasites replicate in a host closely related to humans.

Enteric Hepatitis Viruses

HAV and HEV cause acute hepatitis and self-limited infection in humans and chimpanzees. Although liver disease may be somewhat milder in chimpanzees, the kinetics and magnitude of virus replication, onset of liver disease, and histopathological changes in the liver are similar to those in HAV-infected humans (Dienstag et al., 1975, 1976). The course of HEV infection in chimpanzees is variable, ranging from low viremia with no obvious liver disease to high viremia with biochemical and histological evidence of hepatitis (Li et al., 2006; McCaustland et al., 2000). This may be similar to the spectrum of disease in HEV-infected humans (McCaustland et al., 2000). HAV and HEV infections are preventable by vaccination. The efficacy of a subunit HEV vaccine was approximately 90 percent in two large human trials in Nepal and China, but there is uncertainty about the durability of protective immunity as currently formulated and how (or if) it will be deployed where needed (Shrestha et al., 2007; Wedemeyer and Pischke, 2011; Zhu et al., 2010). Thus it is likely that endemic and epidemic HEV will remain a cause of serious liver disease in developing countries (Aggarwal, 2011). HEV immunity and pathogenesis are still very poorly understood (Aggarwal, 2011). For HAV, socioeconomic development accompanied by improved sanitation and opportunity for vaccination has changed epidemiology in regions where the virus is still endemic, as illustrated by a recent outbreak in South Korea (Kim and Lee, 2010; Kwon, 2009). Under these circumstances, HAV infection shifts from the first to the second and third decades of life with an associated increase in the severity of disease. This situation has highlighted a gap in knowledge about mechanisms of immunity and hepatocellular injury caused by HAV. Very recent studies in chimpanzees provided insight into patterns of innate immunity and host gene expression immediately after infection with HAV and HEV, with the goal of understanding the pathogenesis of these infections and how they compare to responses elicited by HCV that often establishes a persistent infection (Lanford et al., 2011; Yu et al., 2010a). Follow up studies of adaptive immunity to these viruses in animals should be anticipated. Similar studies in humans will be difficult, if not impossible, because infections with these small RNA viruses are often not symptomatic for several weeks and access to liver may be challenging as there is typically no medical need for liver biopsy.

HBV Worldwide, approximately 500 million people are infected with HBV. Hepadnaviruses are widespread in nature and chimpanzees do harbor indigenous strains of HBV that can be distinguished from human viruses based on genomic signatures despite overall identity of about 90 percent (Barker et al., 1975a, 1975b; Dienstag et al., 1976; Guidotti et al., 1999; Hu et al., 2000; Rizzetto et al., 1981). Chimpanzees are nevertheless highly susceptible to challenge with human HBV. Chimpanzees develop persistent and resolved infections after challenge with the virus (Barker et al., 1975a, 1975b). The incubation period preceding symptoms is long and biochemical evidence of acute hepatitis is associated with parenchymal inflammation, as in man. The magnitude and general pattern of viremia and antigenemia during the acute and chronic phases of infection are also similar between the species (Barker et al., 1975a, 1975b; Kwon and Lok, 2011). Severe progressive hepatitis and cirrhosis observed in some humans appears to be uncommon in chimpanzees. Implementation of universal HBV immunization will gradually reduce the number of human infections in future decades, but there is a current need for therapies to control this chronic condition. Nucleoside analog inhibitors of the HBV polymerase that suppress production of infectious virus have been available for years but do not cure infection. Because the HBV genome cannot be eradicated from the liver, most individuals require life-long therapy (Kwon and Lok, 2011). The problem of HBV-resistance to direct-acting antivirals is increasing, and will probably accelerate in regions where treatment practice and availability of high-quality pharmaceuticals of required potency are inadequate (Kwon and Lok, 2011). Immunotherapy to reactivate effective immunity against HBV is an alternative (Rijckborst et al., 2011). A finite course of type I interferon can reverse immune tolerance in about 30 percent of chronically infected patients, conferring long-term control of HBV infection (Rijckborst et al., 2011). As discussed below, new and perhaps more effective approaches to reverse immune tolerance are being considered for chronic viral hepatitis.

Contemporary research in HBV-infected chimpanzees addresses questions that are highly translational to humans. As for HCV, the animals provide a way to develop titered pools of monoclonal HBV for vaccine and related studies (Asabe et al., 2009). These animals were also used to determine if resistance mutations that arise during antiviral therapy facilitate escape from vaccine protection (Kamili et al., 2009). In this study, chimpanzees vaccinated with a commercial HBV vaccine were challenged with a virus containing mutations in key neutralization epitopes of HBsAg caused by development of lamivudine in the viral polymerase gene that is encoded in an overlapping but alternate reading frame (Kamili et al., 2009). Lamuvidine-resistant variants now circulate in some human populations, so this experiment addressed an important public health problem.

HCV   For HCV, there is no vaccine to prevent infection and therapies remain inadequate despite recent progress in developing direct-acting antivirals that target key replicative enzymes of the virus. It should be emphasized that acute infections sometimes resolve spontaneously and chronic infections can be cured. For these reasons, vaccination to prevent persistence is more realistic than for HIV, even though both viruses present similar challenges in their adaptability to immune pressure. Similarly, the goal of effective therapy is to eradicate the virus rather than simply control chronic infection as in HIV and HBV. Chimpanzees are the only species other than humans with known susceptibility to HCV infection. No chimpanzee homolog of HCV has been found and closely related viruses that consistently cause a similar pattern of resolving and persistent infection have not been described in other species. Only chimpanzees have the correct combination of four entry receptors and other cellular co-factors required to recapitulate key features of human infection. Woodchucks and old- and new-world monkeys tested to date are not susceptible to infection (Bukh et al., 2001). HCV infection of a tree shrew (*Tupaia belangeri*) has been reported, but viremia was intermittent and several orders of magnitude less than that measured in chimpanzees and humans (Amako et al., 2010). HCV infection can either resolve spontaneously or persist in humans and chimpanzees. The typical pattern of virus replication is identical, with high levels of viremia for at least 7-12 weeks followed by a decline that is usually associated with a spike in serum transaminases (Abe et al., 1992; Thimme et al., 2002; Walker, 2010). Virus can fluctuate at low levels for several weeks or months before the infection resolves or persists (Walker, 2010). One study reported a lower rate of chronic infection in chimpanzees (40 percent) than humans (60-70 percent) (Bassett et al., 1998), although there is not unanimity on this point. It is difficult in a retrospective chart review to exclude the possibility that some animals thought to be naïve at the time of HCV challenge were already immune because of prior exposure to unscreened human blood products. At least some of the difference may also be explained by over-estimation of virus persistence in humans because acute resolving infections that are clinically silent or mild are missed. Differences related to the young of age of infection in most chimpanzees, and the dose or strain of virus used for experimental challenge, are also possible. If the rate of persistence is lower, it has had no apparent impact on interpretation of studies on immunity to HCV in chimpanzees, or relevance of the findings to human infection (reviewed below). Liver disease is typically mild in persistently infected chimpanzees, as it is in most humans with chronic hepatitis C. More serious liver disease may become evident after several decades. Late-stage disease, including hepatocellular carcinoma, has been observed in some animals more than 30 years after HCV or HBV infection. It should be emphasized that sub-clinical hepatitis over a course of many years falls within the spectrum of hepatitis observed in many humans, and is not uncommon in those without risk factors for rapid progression like male sex, older age, and alcohol intake. Objectives of current chimpanzee research are directly relevant to human health. Highly translatable studies include the first evidence that interferon-free control of HCV infection is possible with combinations of direct-acting antivirals (Olsen et al., 2011), and that interference with a cellular microRNA can prevent HCV replication (Lanford et al., 2010). The latter represents an entirely new approach to control of virus infections. More basic studies focused on the balance between innate and adaptive immunity in HCV infection outcome (Barth et al., 2011), and patterns of host gene expression in liver before infection is clinically evident in humans (Yu et al., 2010b).

In summary, whether chimpanzees are required for progress in understanding and controlling human infectious diseases is highly dependent on several factors, including the availability of valid alternatives, intricacies of the relationship between the host and each pathogen, and the objective of the research. Here, malaria, RSV, HIV and the hepatitis viruses were used as case studies to illustrate these points. To summarize:

- Malaria provides an example where there are alternatives to chimpanzee research, including experimental infection of humans, lower primates, and rodents. The animal model may retain value for testing new vaccine concepts, identification of candidate antigens, and characterization intrahepatic immunity, especially if current strategies to protect humans from infection are inadequate.

- For other pathogens like RSV, the chimpanzee provides the only faithful model of human disease even though lower species are permissive for infection. This has been critical for development of candidate RSV vaccines that have the potential to cause harm.

- The example of HIV illustrates how an animal model can fail despite a very close genetic relationship to humans. Adaptation of SIVcpz to humans after it crossed the species barrier apparently attenuated its replication and pathogenicity for chimpanzees, the species from which it originated. A second very important point that emerged from the HIV experience is that the objectives of vaccination will determine utility of the animal model. If the goal of vaccination is to prevent serious progressive disease (rather than infection), it must be carefully balanced against ethical considerations and any scientific limitations of the model.

- Hepatitis virus research in the chimpanzee has a track record of success that began almost half a century ago. It continues to the present day. As described in more detail later in this paper, proof that humoral and/or cellular immunity protect against HCV infection was generated from chimpanzee studies before the initiation of vaccine trials in humans. As an example, chimpanzees studies within the past decade documented the critical need for T lymphocytes to control HCV infection (Grakoui et al., 2003; Shoukry et al., 2003), and that a vaccine based on this principle could dramatically alter primary viremia (Folgori et al., 2006). These experiments were direct antecedents of current human clinical trials (ClinicalTrials.gov NCT01070407). Publications within the last 18 months addressed important public health concerns surrounding HBV escape variants and vaccination, and tested new concepts for control of chronic hepatitis C virus infection. As noted above, liver disease caused by chronic hepatitis B and C is usually at the mild end of the spectrum observed in humans, but to date this has not been a barrier to successful development of vaccines or therapeutics that target the viruses.

Comparative and Evolutionary Immunology in Humans and Chimpanzees

Humans and chimpanzees diverged approximately 5-7 million years ago. In the 19th century paleontology and comparative anatomy were used to study kinship between the species, but the advent of serology provided a new avenue for investigation. Landsteiner and Miller summarized these studies in a 1925 publication that documented distinct differences between humans, chimpanzees, and orangutans in patterns of hemagglutination by anti-erythrocyte sera (Landsteiner and Miller, 1925). It was nonetheless concluded that chimpanzees and humans were closely related as serology revealed that "the arthropoid apes do not rank in the genealogical tree between lower monkeys and man." Evolution of the immune system remains an important approach to probe the relationship between the species. Publication of the draft genome sequence of a common chimpanzee (*Pan troglodytes*) 80 years after the Landsteiner study facilitated a comparative analysis with the human genome (Mikkelsen et al., 2005). The genomes diverged by approximately 1 percent when estimated polymorphism was excluded. A total of 13,454 pairs of human and chimpanzee genes with unambiguous 1:1 orthology were identified (Mikkelsen et al., 2005). Alignment revealed that the most rapidly diverging gene clusters in both species were associated with taste, olfaction, reproduction, and immunity. With regard to immunity, rapid diversification of chemokine ligands, cytokine biosynthesis, human leukocyte antigen (HLA), and immunoglobulin-like receptors could be discerned even at the relatively close evolutionary distance of human-chimpanzee divergence. Over the past 30 years much has been learned about chimpanzee and human immunogenetics through studies of rapidly evolving genes in the major histocompatibility complex (MHC) and killer cell immunoglobulin-like receptor (KIR) family. Similarities and differences in genes that regulate immunity are reviewed in this section. How these studies of gene evolution in chimpanzees have facilitated and provided insight into infectious disease research is also discussed.

The Major Histocompatibility Complex

Immunogenetic differences between humans and chimpanzees were first explored in the 1960s when mixed leukocyte cultures (Bach et al., 1972) and isoantisera (Balner et al., 1967) were used to define antigenicity of chimpanzee leukocytes. This interest in transplantation biology led to initial characterization of the chimpanzee histocompatibility complex that is now designated Patr (*Pan troglodytes*). During this era experimental transplantation of chimpanzee liver to pediatric patients suffering from biliary artersia was undertaken (Giles et al., 1970). Contemporary immunogenetic research involving the chimpanzee has focused on the evolutionary relationship with man (Lienert and Parham, 1996). The HLA and Patr gene complexes are remarkably similar considering the genetic polymorphism at class I and II loci (Lienert and Parham, 1996). Humans and chimpanzees have orthologous MHC class I A, B, and C loci. Remarkably, there are no species-defining characteristics amongst the highly polymorphic alleles at these loci (Lienert and Parham, 1996). For instance, it is not possible to distinguish HLA-A from Patr-A alleles based on genetic signature. Class II loci are similarly conserved, as humans and chimpanzees express DP, DQ, and DR gene products (de Groot et al., 2009). Chimpanzee and human class I genes are functionally identical. Detailed studies of peptide binding to chimpanzee and human class I molecules demonstrated remarkable overlap in the pool of viral epitopes presented to T cells (Mizukoshi et al., 2002; Sidney et al., 2006). As noted below, complete characterization of Patr haplotype in virus-infected chimpanzees has facilitated adaptation of state-of-the-art reagents for monitoring T cell immunity. Finally, specific class I MHC alleles have been associated with infection outcome in HIV- and HCV-infected humans. Some of these protective alleles appear to have functional orthologs in the chimpanzee, as a subset of Patr class I alleles were shown to bind highly conserved HIV gag epitopes associated with protection from AIDS (de Groot et al., 2010).

The KIR Gene Family

Natural killer (NK) cells are involved in regulation of pregnancy and host defense. With regard to host defense, cytolytic activity and production of effector cytokines by NK cells is tightly controlled by the interaction of activating and inhibitory receptors with their ligands, the class I MHC molecules (Jamil and Khakoo, 2011; Lienert and Parham, 1996; Parham, 2008). Comparison of human and chimpanzee class I ligands over the past 3 decades has stimulated recent interest in evolution of NK receptors that might influence the outcome of infection. Two primary groups of NK receptors have been described. The NKG2 family is relatively non-polymorphic and conserved between humans and chimpanzees. The other family, comprised of KIRs, is highly polymorphic and rapidly evolving as observed in the draft genome sequence of the chimpanzee (Mikkelsen et al., 2005). Humans and chimpanzees each have 10 variable KIR genes but only two, designated 2DL5 and 2DS4, are common between the species. In humans, but not chimpanzees, KIR genes are organized into two haplotypes proposed to roughly correlate with host defense (haplotype A) and reproduction (haplotype B) functions (Abi-Rached et al., 2010). Evolution may have also altered the functional profile of human KIR gene products, as species-specific mutations that reduce avidity of activating KIR for HLA class I, while retaining high-avidity inhibitory KIR, have been found (Abi-Rached et al., 2010). It is apparent that these evolutionary changes over the past 7 million years were driven by selection pressure from infectious diseases and possibly the physiological demands of reproduction in humans versus chimpanzees.

KIR gene diversity between the species may influence the outcome of chimpanzee infections with human pathogens. Resolution of human HCV infections has been associated with homozygous expression of the KIR2DL3 receptor and its specific HLA-C ligand (Khakoo et al., 2004). A KIR haplotype association with HBV infection, and a specific protective effect of KIR2DL3, has also been reported (Gao et al., 2010). It is important to emphasize that these associations were identified in large population studies and the effect is not sufficiently strong to have practical predictive value for individuals. Overall, inhibitory and activating functions of the KIR genes are conserved in humans and chimpanzees. Given the complexity and redundancy of compound KIR:HLA genotypes on NK responsiveness, it is unlikely that KIR genetics have a material impact on typical studies of immunity or vaccine protection involving small populations of humans or chimpanzees.

T Cell Receptor Genetics

T cell recognition is mediated by the heterdimeric T cell αβ receptor (TCRαβ) that recognizes antigens presented by MHC class I or II complexes. TcRβ chain diversity is generated by the rearrangement of V, D, J, and C regions. The random insertion of non-germline-encoded nucleotides at the junctions of these rearranged segments provides additional diversity and is the main site of Ag recognition (complementarity determining region [CDR3]). The human TcRVβ repertoire consists of 54 functional TcRVβ genes belonging to approximately 25 families based on DNA sequence similarities. Partial characterization of the chimpanzee TCR repertoire revealed 42 TcRVβ genes that could be aligned with known human genes (Jaeger et al., 1998; Meyer-Olson et al., 2003, 2004). All functionally rearranged human TcRVβ families were represented in the chimpanzee TcRVβ repertoire. No evidence of new TcRVβ families was found in the chimpanzee, and some genes were identical between the species (Meyer-Olson et al., 2003, 2004). These data indicate a high degree conservation of the TcRVβ repertoire in humans and chimpanzees, and suggest complexity of the T cell repertoire responding to highly mutable viruses like HCV is similar.

Innate Immunogenetics

Innate immune defenses are a potentially important determinant of infection outcome in humans and chimpanzees. This was recently documented in human HCV infection, where a chronic outcome of infection and response to therapy was strongly influenced by a polymorphism in the non-coding region of the IL-28β gene. Whether the same IL-28β polymorphisms exist in the chimpanzee is not yet known, but seems likely given that the range of infection outcomes is identical with humans. Most information on coding sequence differences in innate genes has derived from comparison of evolutionary pressures on key gene families since separation of humans and chimpanzees from a common ancestor

(Barreiro and Quintana-Murci, 2010). These studies have provided insight into natural selection of human and chimpanzee defense genes by infectious diseases. Loss of genes that regulate innate immunity from the chimpanzee but not the human genome has been described. For instance, three genes (IL1F7, IL1F8, and ICEBERG) that appear to be deleted from the chimpanzee genome are involved with regulation of pro-inflammatory responses. ICEBERG is an inhibitor of IL-1β and its loss may indicate a species-specific modulation of inflammasome function, perhaps to reduce sepsis risk (Mikkelsen et al., 2005). Relatively little is known about how deletions or even coding sequence differences in innate genes (which are usually minor) alter immune responsiveness in these species. Using the toll-like receptors as an example, natural selection studies have documented that six chimpanzee TLR genes fit within the range of haplotypes found in European-American, African-American, and Indian human populations (Mukherjee et al., 2009). Despite this similarity, the modal human haplotypes are many mutational steps away from the chimpanzee haplotypes indicating species-specific adaptation to pathogens (Mukherjee et al., 2009). From a practical standpoint, the TLR genes from both species are very close in sequence. For instance, chimpanzee and human TLR4 gene sequences differ at only three amino positions (Smirnova et al., 2000). Differences in patterns of gene expression were observed in primary monocytes stimulated with the TLR agonist lipopolysaccharide (Barreiro et al., 2010). A difference in the number of responding genes in human (335) versus chimpanzee (273) monocytes was observed. Many of the activated genes common to both species were regulated by the transcription factor NFκB and involved in host defense (Barreiro et al., 2010). Others were species-specific, and fell into gene families related to apoptosis (for humans) or SIV control (chimpanzees) (Barreiro et al., 2010). The impact of differences in innate gene coding sequences to the study of specific pathogens in the chimpanzee is unclear, but might be greatest for viruses that originated in the animals and adapted to a new human host. As noted above, HIV strains that adapted to interfere with human tetherin may lose their ability to replicate efficiently in chimpanzee $CD4^+$ T cells, or to infect via the CCR5 co-receptor because of species-specific differences in regulation of gene expression (Wooding et al., 2005).

In summary, comparison of the chimpanzee and human genomes has revealed remarkable conservation of genes; about 30 percent are identical and single base pair substitutions account for about half of the genetic change (Mikkelsen et al., 2005). At the same time, selective pressure against genes associated with immunity is apparent, and almost certainly attributable to infectious diseases that uniquely afflict each species. It is likely that most of these coding differences have limited impact on the value of the chimpanzee as a model for most infectious diseases because of functional redundancies common to immune pathways. An important exception may be a virus like HIV that targets the immune system and only recently adapted to humans after zoonotic transmission from chimpanzees, the intended animal model.

The high degree of protein sequence homology between the species has practical significance for studies of immunity and evaluation of therapeutics like monoclonal antibodies. A body of published literature has documented that most antibodies against cluster of differentiation (CD) antigens that define lymphocyte subsets, differentiation status, and function are fully cross-reactive for human and chimpanzee mononuclear cells. Most importantly, some of these molecules (and others not associated with immunity) are considered targets for monoclonal antibody therapy of human diseases. Examples are provided below. Advantages of chimpanzees as a pre-clinical model for monoclonal antibody development have been summarized elsewhere (VandeBerg et al., 2006), but include increased probability of detecting unintended effects against proteins that are orthologous to the primary target, similar binding affinities that might alter cellular responses to a therapeutic antibody, and identical pharmacokinetics of human and humanized antibodies in humans and chimpanzees but not lower primate species.

Functional Immunology and Vaccine Research in Humans and Chimpanzees

Evolutionary studies have revealed similarities and differences in immune response genes between the species. How different coding sequences in immune response genes alters infection and immunity in a chimpanzee versus a human is difficult to predict and probably pathogen-specific. Infection of chimpanzees with the hepatitis C virus illustrates the strengths and limitations of studying immunity to a human virus in chimpanzees. Non-A, non-B hepatitis (hepatitis C) was first studied in chimpanzees to ask fundamentally important questions unrelated to immunity. For instance, the ability to transmit non-A, non-B hepatitis from humans to chimpanzees indicated an infectious etiology of disease. It also facilitated physico-chemical characterization of the agent as a small, enveloped RNA virus and provided a pedigreed stock of infectious serum for molecular cloning of the HCV genome as noted above. The observation that some animals, like humans, developed chronic hepatitis C while others spontaneously cleared the virus provided a unique opportunity to identify protective immune responses that might be relevant to humans. In this section, functional adaptive immune responses elicited by infection with HBV and HCV in chimpanzees and humans are compared. Value of the animals for vaccine development is also highlighted. Limitations of the model, and examples of experimental approaches that can be taken in chimpanzees but not humans are described.

Adaptive CellularIimmunity

Detailed studies of cellular immunity to HBV and HCV have provided insight into mechanisms of protection from persistence and how these responses fail.

Chimpanzees offer at least three distinct advantages for this research: (1) The first few weeks of HBV and HCV infection are clinically silent and so critical events that shape the adaptive immune response and infection outcome are very difficult to study in humans; (2) animals can be challenged with well-defined HCV quasispecies and even clonal HCV genomes to facilitate studies of virus adaptation to the host and immune selection pressure; and (3) the liver can be sampled by percutaneous needle biopsy from the earliest times after infection, so that patterns of innate and adaptive gene expression can be studied.

It is important to emphasize that the tools for measuring cellular immunity in chimpanzees are as sophisticated as those available for human studies. Antibodies to key differentiation, regulatory, and effector molecules expressed by human T cells cross-react with the equivalent chimpanzee molecules. Virus-specific T cell responses in chimpanzees can be very precisely quantified by functional assays that measure production of effector cytokines or killing. As noted above, evolutionary studies of the chimpanzee Patr complex provided insight into MHC class I and II restriction of the T cell response to hepatitis viruses in chimpanzees. This work on the Patr complex also facilitated development of soluble class I and II molecules (tetramers) for direct visualization of virus-specific T cells in the blood and liver of chimpanzees infected with HCV and HBV. These chimpanzee reagents are produced by an NIH-funded facility that established to provide human and murine class I and II tetramers. In summary, there are no technical disadvantages, and several distinct advantages, to the study of antiviral T cell immunity in chimpanzees versus humans.

Patterns of virus replication and T cell immunity are identical in humans and chimpanzees infected with HCV and HBV (Cooper et al., 1996; Guidotti et al., 1999; Rehermann, 2009; Thimme et al., 2001, 2002, 2003; Walker, 2010). For instance, HCV replicates at high levels for 8-12 weeks before the onset of $CD4^+$ helper and $CD8^+$ cytotoxic T cell responses that are associated with a spike in biochemical markers of hepatitis and initial control of viremia (Rehermann, 2009; Walker, 2010). In some humans and chimpanzees, the T cell response is sustained and the infection is terminated within a few days or weeks. In others, the $CD4^+$ T cell response fails and the virus persists. Failure of the HCV-specific $CD4^+$ T cell response before apparent resolution of infection is the best predicator of a chronic course of infection. Importantly, mechanisms of acute phase $CD4^+$ T cell failure remain unknown (Rehermann, 2009; Walker, 2010). HCV-specific $CD8^+$ T cells are present in the liver at high frequency for decades but provide no apparent control of virus replication. Infection of chimpanzees with viruses of known sequence was essential to show that some of the HCV epitopes targeted by these $CD8^+$ T cells acquired escape mutations that prevent recognition of infected cells (Bowen and Walker, 2005). Other long-lived intrahepatic $CD8^+$ T cells target intact epitopes but lack effector functions (Rehermann, 2009; Walker, 2010). Based on these observations, current research in chimpanzees has two goals. The first goal is to facilitate development of vaccines that skew the outcome of HCV infection from persistence to resolution. The second goal is to determine the defect that underlies $CD4^+$ and $CD8^+$ T cell failure in chronic hepatitis C (and B), and to test approaches to reverse exhaustion.

Vaccine Research

Very early studies demonstrated that spontaneous resolution of HCV infection in a chimpanzee did not protect from liver disease when the animal was re-exposed to the same infectious inoculum (Farci et al., 1992; Prince et al., 1992). It was concluded that anti-HCV immunity was weak and that vaccine development would be difficult. Subsequent studies revealed that this was not necessarily the case. While some second infections in naturally immune animals do persist, the majority of infections are very rapidly controlled (Walker, 2010). As noted above, primary HCV infections typically do not resolve for 3-4 months, but most second infections clear within days, and are associated with an accelerated memory T cell response. The chimpanzee model was essential to prove the importance of memory $CD4^+$ and $CD8^+$ T cells to protection from persistence. Animals that had successfully resolved two infections were treated with monoclonal antibodies directed against CD4 or CD8 to temporarily deplete these subsets before a third challenge with HCV. In the absence of $CD4^+$ T cells, the virus persisted and $CD8^+$ T cells (that were not depleted) selected for virus variants with escape mutations in class I epitopes (Grakoui et al., 2003). Depletion of $CD8^+$ T cells in a second set of immune chimpanzees prolonged a subsequent infection; termination of the infection coincided with recovery of these effector cells (Shoukry et al., 2003).

Together, these studies demonstrated that sterilizing immunity provided by antibodies is not necessarily required for HCV protection. Instead, they indicated that induction of T cell immunity to prevent persistence (but not infection) may be a realistic goal for vaccination. Indeed, the T cell depletion studies in chimpanzees led directly to the design of a recombinant adenovirus vector that expressed the non-structural proteins of HCV that are predominantly targeted by the cellular immune response (Folgori et al., 2006). Structural proteins, including envelope glycoproteins that are the targets of neutralizing antibodies, were not incorporated into the vaccine. Chimpanzees vaccinated with this vector had dramatically lower levels of primary HCV viremia than mock-vaccinated controls, and all cleared the infection (Folgori et al., 2006). This vaccine is now in human clinical testing for prevention and treatment of HCV infection (see ClinicalTrials.gov NCT01070407, NCT01094873, and NCT01296451). It should be noted that these vaccines developed in Europe by IRBM (Merck) and Okairos were evaluated in chimpanzees at primate facilities in the United States.

The contribution of antibodies to vaccine-mediated protection against HCV also cannot be minimized, as indicated by a recent meta-analysis of chimpanzee vaccine studies (Dahari et al., 2010). Active vaccination with recombinant subunit vaccines comprised of the HCV E2 envelope glycoprotein protected chimpanzees from virus challenge. Proof of protection by these vaccines was obtained in chimpanzees before initiation of phase I and II testing in humans. Finally, chimpanzees were used to test post-exposure prophylaxis with anti-HCV antibodies, a potentially important approach to infection control in the setting of needle-stick injury (Krawczynski et al., 1996). These antibodies substantially modified the course of acute hepatitis C, but did not prevent persistence of the virus.

Therapeutics

The last decade has brought tremendous progress in understanding mechanisms of T cell evasion by persistent viruses. Studies in murine models of virus (LCMV) persistence have demonstrated that functionally impaired virus-specific T cells express the inhibitory molecule PD-1. Antibody-mediated interruption of the PD-1 interaction with its ligand at least partially restores T cell function and leads to accelerated control of virus replication (Barber et al., 2004). It is now clear that exhausted T cells in humans and chimpanzees persistently infected with HCV and HBV express multiple inhibitory receptors including PD-1 that put a brake on effector function (Boni et al., 2007; Golden-Mason et al., 2007; Penna et al., 2007; Raziorrouh et al., 2010). Interruption of ligand binding by inhibitory PD-1, CTLA-4, and TIM-3 receptors on HCV-specific $CD8^+$ T cells restores function in cell culture assays (Boni et al., 2007; Golden-Mason et al., 2007; Penna et al., 2007; Raziorrouh et al., 2010). Initial studies in persistently infected chimpanzees indicate that PD-1 blockade can have a dramatic effect on viremia in some but not all animals (C. Walker, unpublished). Similar studies have been completed in humans (ClinicalTrials.gov NCT00703469) but the results have not been released. Based on studies in the animal model, it might be predicted that the human trial had some successes but more failures.

There is a continuing need for chimpanzees to develop next-generation therapeutics against persistent viruses, especially those that cannot be eradicated from infected cells (Callendret and Walker, 2011). Very recent cell culture studies have indicated that blockade of one inhibitory receptor may not be adequate to fully restore function the HBV or HCV specific T cells (McMahan et al., 2010; Nakamoto et al., 2009). Blockade of multiple pathways is feasible, but the approach carries risk as these pathways were designed to temper unwanted or dangerous immune responses (Callendret and Walker, 2011). Also, studies in mice indicate blockade of one of more than inhibitory pathway in combination with vaccination might potentiate activity against persistent viruses (Ha et al., 2008). In considering these strategies for human use, two considerations are paramount:

(1) Most humans with chronic hepatitis are relatively healthy and so combinations of blocking antibodies and vaccines carry more risk than studies in patients with advanced stages of cancer. Under these circumstances, a nonhuman primate model like the chimpanzee is important for progress.

(2) Antiviral therapy for chronic hepatitis C will become more effective with the advent of small molecules antivirals. It is too soon to know if there is a place for immunotherapy with vaccines and/or blocking antibodies in hepatitis C, although the propensity of the virus to develop resistance is remarkable and a perhaps a significant barrier to availability and use of the multiple drug cocktails in developing countries. There is a significant need for this type of therapy in chronic hepatitis B, where direct-acting antivirals do not eradicate the infection and it is necessary to break immune tolerance for long-term control of virus replication. Chimpanzees will remain important to advance this concept.

The Chimpanzee in an Era of New and Emerging Technologies for Studying Immunity

Chimpanzee infection studies continue to have value beyond proof-of-concept studies to validate new preventive or therapeutic strategies. New technologies offer great promise in unraveling molecular mechanisms underlying the failure of immunity in acute and persistent infections with viruses like HAV, HBV, and HCV. It is now possible to probe the innate and adaptive immune responses, and how they are coordinated, a level of resolution not possible a few short years ago. Genes that are expressed in virus-specific T cells that successfully control infection or become exhausted can be identified with new technologies (Haining and Wherry, 2010). Chimpanzees may be integral to the future of this work that could provide new targets for intervention in acute and persistent infections. Human genomic and proteomic technologies are directly adaptable to the chimpanzee, and isolation of antiviral T cells to high purity is possible. Perhaps most importantly, the chimpanzees provide unique access to paired blood and liver specimens at very early time points after virus exposure when there are no symptoms but the outcome of infection is probably determined. It is likely that molecular characterization of antiviral T cells in liver at the earliest stages of infection will identify and validate therapeutic targets relevant to humans.

MACAQUES AS MODELS FOR HUMAN DISEASE

In this section, we address host factors in macaque models for human diseases, macaque immunogenetics, and understanding the roles of innate and adaptive responses in macaque models for the development of vaccines and immunotherapies. With 93 percent sequence identity with humans, *Macaca species* that predominate in Asia are the most widely utilized nonhuman primates in biomedical research. The three most commonly utilized in biomedical research today are *M. mulatta* (rhesus), *M. nemestrina* (pigtailed) and *M. fascicularis* (crab-eating or longtailed); and within the genus the rhesus macaque is by far the most frequently studied. The landmark work by Landsteiner and Wiener in 1937 to define Rh factors in blood, allowing blood typing, was an early example of the contributions that this species has made to medical research (Landsteiner and Wiener, 1937, 1941). Efforts since then have been focused on developing multiple models for understanding human disease states. Macaques share many similarities with humans and chimpanzees in their hematology, reproductive biology, neurological development, behavior, immunogenetics, and immune responses to pathogens. Much of this subsequent research has focused on models in the rhesus macaque, due to their relative ease of breeding in captivity and high adaptability to novel environments. These models include (references are representative examples and not comprehensive in nature): reproduction, including stem cell research (Ben-Yehudah et al., 2010; Schatten and Mitalipov, 2009; Tachibana et al., 2009), bone marrow transplantation and hematopoietic stem cell gene therapy (Donahue and Dunbar, 2001), aging (Messaoudi et al., 2006), metabolic diseases and their sequelae such as diabetes (Grove et al., 2005), brain and neurological development (Sarma et al., 2010; Soderstrom et al., 2006; Voytko and Tinkler, 2004), behavioral (Bethea et al., 2004; Sabatini et al., 2007; Stevens et al., 2009) including addiction (Barr et al., 2010), and infectious diseases (Daniel et al., 1985; Haigwood, 2009; Hansen et al., 2010; Messaoudi et al., 2009) including virus-associated malignancies (Messaoudi et al., 2008). There has also been some work to understand autoimmunity and arthritis (collagen-induced arthritis and spontaneous arthritis) in macaques, reviewed in Vierboom and 't Hart, 2008. With the advent of array technologies to examine multiplex responses to disease, it will be increasingly possible to identify the roles of innate and adaptive immunity in the macaque models for human disease models. Models for aging and immune senescence have been developed using rhesus macaques that are 18 years and older. These animals are characterized by a progressive loss of naïve T cells and an accumulation of memory type T cells with age. The models can be used to examine therapeutic approaches to reinstating naïve T cells, such as IL-7 therapy (Aspinall et al., 2007) and caloric restriction (Messaoudi et al., 2006). At this time, one of the best probes for understanding host immunity is to utilize pathogens that elicit similar responses to infection in humans and macaques. To introduce this subject, we provide a brief review of host restriction factors and infectious disease models for human disease in the macaque.

Host Factors in Macaque Models for Human Diseases

Host Interactions with Pathogens at the Cellular Level

A major distinction for macaques compared with chimpanzees is that their greater genetic distance from humans results in lack of susceptibility to certain human pathogens. Viral agents and intracellular pathogens utilize host cellular receptors and intracellular molecules such as the nucleic acid polymerases and transcription machinery for replication and propagation. In order to assure priority treatment when inside the cells, these organisms or viruses utilize specific proteins encoded in their genomes to interfere with basic cellular activities such as host protein production. In contrast, extracellular bacterial and protozoan parasites (at least in some stages of their life cycles) replicate independently of the human cell and thus are more capable of establishing infection in a wider range of hosts. Typically pathogens are adapted to certain hosts, a property termed "host range" that is conferred by a number of factors. For viruses, it has become clear that there are specific restriction elements that interact with portions of the virus to limit replication or assembly. Concomitantly, viral proteins can confer resistance to restriction elements in "spy versus spy" interplay at the molecular level, such as Nef from HIV-1 or SIV (Schmokel et al., 2009). No fewer than four different host-pathogen mechanisms have been identified to limit the cytopathic effects of HIV and SIV alone, summarized in a recent review (Lifson and Haigwood, in press) and publication on the latest factor to be discovered (Laguette et al., 2011; St Gelais and Wu, 2011). Generally speaking, viruses with larger genomes such as those from the herpesvirus family are better equipped to stave off the antiviral effects of the host, with multiple pathways for downregulation of various host proteins, including those that are critical for immune responses such as MHC Class I proteins (Hansen et al., 2010). The process of adaptation of viruses during zoonoses is one of accruing sufficient mutations to overcome these host factors, and one of the best examples of recent zoonoses that have had a major impact upon human health is the transmission of HIV-1 and HIV-2 to humans (Gao et al., 1999; Hahn et al., 2000). A well-adapted virus is one that can peacefully coexist with the host in the absence of pathogenic effects, an example being SIV in African nonhuman primates (sooty mangabeys, African green monkeys, and mandrills) (Pandrea et al., 2006; Sodora et al., 2009). Understanding the immunological differences between pathogens that are recently acquired and poorly adapted, compared with those that are established and benign, may yield important information about both the host and the virus (Silvestri, 2009). The consequence of this host range restriction is that certain types of infectious disease and immune-based research can only be performed in chimpanzees or in humans, and not in macaques. In this section, we provide examples of macaque models for major human diseases that utilize either the same microbe or species-specific pathogens that are related to the human pathogens.

Pathogenic Models for Human Diseases Using the Same Organism

Due to the physiological and genetic similarities of humans and macaques, many human pathogens can infect and cause disease in macaques, and the degree of adoption for experimental usage is dependent upon several factors such as length and severity of the disease, as well as similar pathologic outcomes to humans. An example is tuberculosis, which is highly infectious in macaques and represents a good model for acute and latent infection as well as re-activation and progressive disease (Chen et al., 2009; Lewinsohn et al., 2006). Other models include acute infections that are typically resolved in humans, such as measles (McChesney et al., 1997; Permar et al., 2007; Zhu et al., 1997), potentially severe acute respiratory syndrome (SARS), (Miyoshi-Akiyama et al., 2011), Ebola (Sullivan et al., 2006), monkeypox (Estep et al., 2011), and dengue virus (Onlamoon et al., 2010; Smits et al., 2010). The considerable work that goes into model development for these agents includes the production and in vivo titration of infectious challenge stocks, in some cases necessitating testing many different sources to find the right balance of infectiousness and pathogenicity, the development of key assays for monitoring disease sequelae and the progression of infection short of necropsy, and the development of reagents that are appropriate for monitoring innate and adaptive immunity in vivo (discussed below). In addition, experiments using these agents require special containment to reduce biohazardous exposure to humans and other nonhuman primates. However, significant progress has been made not only in understanding correlates of protection from challenge in some cases but also in developing human vaccines that show some promise, discussed with examples in the section below on vaccines. Obvious advantages of these models include the ability to identify virulence genes and to perform more frequent and more invasive sampling, with the judicious use of serial sacrifice studies to examine tissues in more depth.

Pathogenic Models for Human Diseases Using Closely Related But Distinct Organisms

When host range limits the infectivity of certain agents, closely related pathogens that are adapted for *Macaca species* can sometimes be found. These include: simian adenoviruses, simian varicella virus (SVV) as a model for varicella zoster virus (Messaoudi et al., 2009), rhesus cytomegalovirus (CMV) as a model for human CMV (Hansen et al., 2010), and SIV and chimeric simian/human immunodeficiency virus (SHIV) as a model for HIV-1 (Lifson and Haigwood, in press). Because the HIV-1 and SIV Envelope proteins do not elicit cross-protective neutralizing antibodies, testing of human monoclonal antibodies or HIV Env-based vaccines for protection was not possible with SIV. Therefore chimeric viruses were made consisting of the SIV backbone and a "swapped" HIV-1 env gene and these viruses were passaged in vivo in macaques to obtain more pathogenic strains. Originally, SHIVs were tropic for CXCR4 and caused rapid loss of CD4+ T cells in the periphery (Reimann et al., 1996), but then CCR5-utilizing isolates with slower pathogenesis were successfully constructed and shown to be transmitted by mucosal routes in adult (Harouse et al., 2001) and newborn macaques (Jayaraman et al., 2007). Despite their appeal, these models have potentially more limitations that must be kept in mind in interpreting the results. The specific agent (genetically or phenotypically characterized), the host, or that particular combination can each contribute to differences in disease outcomes. All of the caveats noted in the section above hold in this case, and in addition there is the potential for difficulty in translating the findings from macaque-specific pathogens to their human homologs. Nonetheless, these models can be and have been highly instructive in establishing certain basic principles that would have been difficult or impossible to determine by experimentation in humans or in chimpanzees, as summarized in Table 1. Although many of these concepts have since been proven in clinical studies, it can be argued that the timing of the discovery of the concept in the macaque accelerated understanding and exploration in human studies. Furthermore, many of the findings could not be directly tested in humans for ethical or safety reasons.

TABLE 1 Examples of Lessons Learned from Macaque Models for Human Diseases

Human Pathogen	Macaque Pathogen	Concept Revealed	References
HIV-1	SIV	Mucosal routes of infection require greater doses than parenteral for productive infection after a single challenge; influenced by hormonal status	(Hirsch and Lifson, 2000; Sodora et al., 1997)
	SIV	Timing of tissue distribution using serial sacrifice; earliest events in infection	(Milush et al., 2004)
	SIV	Multiple low dose mucosal challenge with can mimic the transmission dose and composition, similar to sexual mucosal acquisition of HIV-1 in humans of a few variants to seed the infection	(Keele et al., 2009b)
	SIV, SHIV	Relatively more rapid pathogenesis in newborns compared with adults	(Baba et al., 1995, Jayaraman et al., 2007; Marthas et al., 1995)
	SIV	Early loss of GALT	(Smit-McBride et al., 1998)
	SIV	Neurotropic strains of SIV and HIV-2	(Van Rompay and Haigwood, 2008)
		Superinfection definitively Demonstrated	(Yeh et al., 2009)
	SIV, SHIV	Differential pathogenesis conferred by the same viruses in different macaque species	(Polacino et al., 2008; Sodora et al., 2009)

Human Pathogen	Macaque Pathogen	Concept Revealed	References
Human CMV		Mechanisms of superinfection revealed	(Hansen et al., 2010)
Ebola virus	Ebola virus	Demonstration that antibodies are a correlate of protection but are not sufficient for protection	(Sullivan et al., 2009)

Macaque Immunogenetics

Due to the more extensive use of the Indian-origin rhesus macaque, this was the first of the macaque genomes to be sequenced. The draft DNA sequence of an Indian-origin female rhesus was completed in 2007 and compared with the human and chimpanzee genomes (Gibbs et al., 2007); the Chinese rhesus macaque genome was just published in 2011 (Fang et al., 2011). This three-way comparison study gave some of the first insight to inform an understanding of mutational mechanisms that have, during the last 25 million years, shaped the biology of the three species. Prior work had focused on specific regions of the genome that encode gene members of the immune system, as noted below.

Major Histocompatibility Complex

MHC loci in rhesus macaques have been explored and compared, with a great deal of progress in the rhesus macaque. Class I alleles are termed Mamu (for *Macaca mulatta*) (Doxiadis et al., 2007; Otting et al., 2007); in contrast to humans only two MHC class I loci are found, A and B, with at least two expressed B loci, indicating a duplication of the B locus. Macaques have three MHC class II loci: DP, DQ, and DR. Haplotype diversity can result from crossing over events, since rhesus macaques have several class I alleles on each chromosome. Comparison of rhesus and human class I and class II evolution shows that the class I alleles are not shared between the species (Boyson et al., 1996), although similar epitope binding motifs are shared by macaque and human MHC class I molecules (Loffredo et al., 2009). MHC class I haplotype can clearly affect disease progression in SIV infection, in that certain MHC haplotypes affect the ability to control viral replication in vivo (Bontrop and Watkins, 2005; Goulder and Watkins, 2008). Mamu-A*01 animals typically show better control of infection in vaccine studies (Pal et al., 2002), and haplotypes B*08 and B*17 are associated with elite control of viremia (Loffredo et al., 2008; Yant et al., 2006). There also appears to be an effect of class I haplotype upon antiretroviral drug effectiveness (Wilson et al., 2008). Certain MHC class I alleles in humans also appear to be associated with better control of viremia and disease outcomes, reviewed in Goulder and Watkins, 2008.

The KIR Gene Family

As noted above in the section on chimpanzees, MHC class I molecules are ligands for the KIRs, which are expressed by natural killer cells and T cells. As with chimpanzee KIRs, the interactions between these molecules contribute to both innate and adaptive immunity, and combinations of MHC class I and KIR variants influence resistance to infections, susceptibility to autoimmune diseases and complications of pregnancy, and outcomes of transplantation (Parham, 2005). The genes encoding the KIRs all arose recently from a single-copy gene during the evolutions of simian primates, after which the KIR and MHC class I genes co-evolved. The genes have been sequenced in rhesus, humans, chimpanzees, gorillas, gibbons, orangutans, and marmosets (Sambrook et al., 2006), and in Mauritian cynomolgous macaques (Bimber et al., 2008). There has been a recent comprehensive rhesus macaque KIR data set developed (Moreland et al., 2011) and an overview of MHC/KIR co-evolution (Parham et al., 2010), emphasizing rapid evolution of KIR sequences, in large part due to evolutionary pressure from infectious pathogens. Of note, the old-world monkeys have been recently described as being the species most likely to provide useful and informative models for human disease.

T Cell Receptor Genetics

The diversity of T cell receptor (TCR) alpha and beta chains is created by somatic recombination of germ-line genes, as described above. Several studies have examined TCRs in chimpanzees and macaques (Jaeger et al., 1993). Macaques that are not infected display a diverse T cell repertoire characterized by a Gaussian distribution of betaCDR3 lengths (Currier et al., 1999). T cell repertoire analysis has revealed a dominance of T cells expressing specific V-beta segments during chronic infection (such as SIV infection). Rhesus CMV infection led to polyclonal CD4+ T cells that changed over time and chronic infection to reveal a skewed hierarchy dominated by two or three clonotypes (Price et al.,

2008). These kinds of comparisons aid in understanding the role of T cells in controlling infection and how these types of changes compare with natural progression of aging, for example (summarized in Messaoudi et al., 2011).

Antibody Gene Families

In macaques as in humans, the diversity of B cell receptors (BCR) and the soluble forms, or antibodies, that result, is generated by somatic recombination as with the TCR. The immunoglobulin loci for antibodies are similarly arranged in the rhesus macaque as in humans, and the antibodies have the same structures, with clear homologs identified for IgG and IgA classes and for subclasses of IgG (Scinicariello et al., 2004). There are differences, as would be anticipated with the time since speciation, but remarkable functional and genetic conservation.

CD Genes

Elegant comparative studies examining the relative proportions and cell surface markers of human versus macaque cells have been described and are summarized in a recent review (Messaoudi et al., 2011). The strong conservation of these molecules has allowed detailed studies on immune cell ontogeny, homing, survival, proliferation, and death that also have opened the field of immune senescence in the macaque. As noted in the sections below, the roles of specific cells in disease have been studied in vivo by transient depletion of subsets.

Innate Immunogenetics

CD16+ and CD56+ NK subsets are largely similar in function and distribution in humans and macaques. The distribution of NK cells in blood and tissues differs somewhat in macaques, where CD16 predominates in the blood, and CD16 negative cells positive for CD56 can be found in tissues (Reeves et al., 2010). Prior studies had suggested that NK cells might not play a strong role in the containment of SIV; the role of NK cells in control of SIV still uncertain but certainly cannot be discounted. Macaques are appropriate models in which to address questions in acute infection, which is a phase of the infection that is very difficult to identify and thus study in humans. There are new data that have identified the macaque counterpart of mucosal NK cells producing IL-22

(Reeves et al., 2011), as previously identified in humans. These studies emphasize the importance of performing studies in macaques where greater access to mucosal samples is possible.

Monitoring and Understanding the Roles of Innate and Adaptive Responses in Macaque Models for the Development of Vaccines and Immunotherapies

Flow cytometric analysis of macaque immune-related cells using murine anti-human CD antibodies that were developed to bind to human cell surface markers demonstrates the strong immunological conservation of the vast majority of these surface molecules. This attribute of shared receptor recognition is not surprising knowing the genetic conservation of the species, and it has meant that the nuances of antigen presentation and the respective roles of B and T cells in macaque immunity is now a well-established area of research. A recent review provides an overview and examples of the staining and separation of these subsets in rhesus macaques (Messaoudi et al., 2011). The relative contribution of specific types of cells in innate and adaptive immune responses in the outbred *Macaca species* has been made possible by infusion studies using murine monoclonals directed at human CD8, CD4, CD16, CD20 (or other cell surface markers of interest) to transiently deplete specific subsets of lymphocytes in vivo. These studies have demonstrated the relative contribution of $CD4^+$ and $CD8^+$ T cells, B cells, and NK cells to the control of specific pathogens at different time points during infection, with examples in the sections that follow.

Stimulating Innate Responses

As in humans, the mediation of innate responses includes neutrophils, NK cells, dendritic cells (DC), and macrophages. Recognition of microbes depends upon the detection of pathogen-associated molecular patterns (PAMPs) by pattern recognition receptors, which fall into two families, toll-like receptors (TLR) and RIG-I-like RNA helicases (RLH). DCs and NKs have been well characterized in the rhesus. The major two subsets, myeloid (mDC) and plasmacytoid (pDC) types, are identified with the same surface markers and share functionality and cytokine response following viral infection. These two types of DCs express the same TLRs as human DCs, which differ drastically from murine DCs, and thus help establish the macaque as an excellent species in which to evaluate TLR ligands as adjuvants (Rhee and Barouch, 2009).

Magnitude and Quality of T Cell Responses

In the T cell realm, there has been significant development work to understand and to quantify specific T cell subsets in the blood and effector sites (Pitcher et al., 2002; Walker et al., 2004), aided by advances in sampling methodologies that provide repeated sampling opportunities at effector sites such as the lung, for example. The functional diversity of the T cell response in macaques, as in humans, can now be measured via "staining" for multiple cytokines simultaneously. Intracellular cytokine staining (ICS) is a commonly utilized technique, where cells are stimulated with antigens (virus, lysates, purified antigens, or peptides) and then treated to block cytokine secretion, then labeled for each cytokine with a different-colored tag, and enumerated on a flow cytometer. Much of this work has been stimulated by interest in HIV pathogenesis and subsequently SIV infections, as these viruses destroy CD4+ T cells in the gut and peripheral blood compartments, with concomitant negative effects on T cell help; the rapidity of destruction varies with the strain. The time course of longitudinal development of T cell help (CD4+ T cells) and cytotoxic cells (CD8[+] T cells) in response to SIV or SHIV infection is similar to that of HIV-1 in humans. In acute SIV infection, CD8[+] T cells were shown to be very important for viral control (Schmitz et al., 1999). More recent studies have further elucidated the impact of CD4 and CD8 T cells, in experiments that clearly show that CD8[+] T cell removal during this timeframe results in rapid disease progression due to unchecked viremia (Okoye et al., 2009); the depletion also resulted in proliferation of CD4 T cells, particularly effector memory cells. Antiviral CD8[+] T cells have also been implicated in controlling vaginal infection upon exposure to a highly virulent SIVmac239 after vaccination with SHIV89.6, an infectious but non-pathogenic strain (Genesca et al., 2009).

Antibodies and Affinity Maturation

The development of antibody responses is dependent upon the effective antigen presentation and, for some diseases, persistence of antigens. Antibody response kinetics are indistinguishable in macaques and in humans, but it has been easier to perform longitudinal studies to examine the maturation of the response with more frequent time points (Cole et al., 1997). Antibodies increase in avidity with time of exposure to antigen and the quality of the response depends upon the form of the antigen. Many persistent pathogens utilize various methods for immune evasion, including antigenic variation of their surface proteins to elude recognition B cells and thus to prevent containment by antibodies. SIV and SHIV have served as excellent models for longitudinal studies of antibody development, including neutralizing antibodies, which typically increase in magnitude and breadth with time (Hirsch et al., 1998; Kraft et al., 2007), as is seen in humans infected with HIV-1 (Mahalanabis et al., 2009; Sather et al., 2009). Antibody responses directed to HIV or SIV Envelope typically are robust and directed to hydrophilic, hypervariable regions; responses to conserved regions (such as those required for receptor and coreceptor binding) arise later and do not appear in all subjects (Li et al., 2007). Variation and post-translational modifications such as N-linked glycosylation in the SIV Envelope protein have been shown to lead to escape (Rudensey et al., 1998). There is now extensive molecular and sequence data of both viral variants and antibodies that HIV and SIV infection leads to similar B cell responses in humans and macaques, respectively. The development of binding and neutralizing antibodies occurs over the same time frame and requires extensive affinity maturation from the germ line antibody genes (Moore et al., 2011). Monoclonal antibodies from SIV-infected macaques have been instructive in understanding the neutralizing epitopes targeted (Cole et al., 2001; Glamann et al., 1998; Robinson et al., 1998), although no extraordinarily powerful monoclonal antibodies have yet been isolated with properties similar to those found in HIV+ subjects who are elite neutralizers (Doria-Rose et al., 2009; Walker et al., 2010; Wu et al., 2010). Contributions by CD20-positive cells in SIV infection were evaluated and shown to be limited to the chronic phase, with no apparent effect on early viremia (Schmitz et al., 1999, 2003). A recent clinical report showed that an HIV-positive subject with lymphoma who was treated with anti-CD20 (Rituximab) to reduce his B cells had a transient increase in plasma virus and a reduction in the level of neutralizing antibodies. These data were interpreted as evidence for contributions to viral control by neutralizing antibodies during the chronic stage of HIV-1 infection, consistent with the data on SIV-infected macaques treated with anti-CD20 (Huang et al., 2010). Depletion studies have been highly informative in other diseases, such as measles (Permar et al., 2003, 2004). A comparison of B cell depletion during the acute phase of infection in this model alone or in combination with CD8[+] T cell depletion clearly showed that the antibodies played a limited role in the control of measles viremia, while the CD8 effector T cells were critically important for limiting viremia and rash production.

Vaccine Approaches in Macaques

Macaque vaccine experiments performed over the last 25 plus years have accrued a large body of data about relative immunogenicity of various vaccine approaches, and these experiments have also allowed correlates of vaccine protection to be determined in many cases. In the sections below, we have attempted to summarize briefly the status of research in five major pathogens, to introduce the major concepts under investigation and the importance of these models to the discovery of effective vaccines for diseases that are emerging, or re-emerging due to persistence in the host, such as tuberculosis. Approaches for vaccines depend upon the desired type of immunity required for protection. Protein formulations are often processed through endocytic pathways that stimulate $CD4^+$ T helper 2 (T_H2) cell responses and promote antibody production. By contrast, vaccines that allow synthesis of foreign proteins within cells lead to processing of antigens through the proteasome, a process that more effectively elicits $CD8^+$ T cell responses, while also eliciting antibody responses. Some gene-based vaccines have the potential to generate broad responses because of their ability to target antigen-presenting cells (APCs) directly, which is a property of certain viral vectors. The quality and range of vaccine-induced immune responses can therefore be influenced by the specificity of viral vectors for different APC targets.

With the advent of molecular tools for genetic manipulation and the identification of virulence factors, it has been possible to utilize this knowledge to build recombinant vectors to precisely excise unwanted genes and to express one or more vaccine antigens in their place. Advances in the development of safer, more attenuated viral vectors utilized for one major disease, smallpox, were the genesis of recombinant poxvirus-based vectors that persist in the human vaccine armamentarium today. These viral vectors are attractive because they stimulate innate and adaptive immunity and persist long enough to provide a strong antigen pool to boost B cells, much as the currently licensed live attenuated vaccines. The ability to challenge in macaques prior to testing in humans provides a measure of confidence that the immune responses elicited by the new vaccines are effective against diseases that closely model human pathologies.

As with model development to study pathogenesis, it is also true that the use of related pathogens as "vectors" for vaccine delivery has also necessitated understanding the host range of the various vectors for humans and for macaques. Thus there has been an intensive search for non-pathogenic viruses that can replicate equally and stimulate similar levels of immunity directed to the foreign antigen(s) in both species. Current investigations with viral vectors include Vesicular stomatitis virus, Sendai virus, adenoviruses (chimpanzee, human, and macaque), poxviruses (various levels of attenuation and species specificity) and herpesviruses (including cytomegalovirus, and adenovirus-associated virus vectors), to name a few. Due to cellular targeting via receptors, replication capacity, and the inclusion of cofactors such as cytokine genes, some vectors are better at stimulating specific arms of the immune response; combinations can work additively and possibly synergistically.

In addition to viral vectors, other promising approaches in use include DNA delivered by a variety of routes (intramuscular, intradermal, via microneedles, and with electroporation to enhance uptake) (Yin et al., 2010). DNA vaccines showed impressive protection in a mouse influenza model (Ulmer et al., 1993) but have been much less immunogenic in macaques and humans with SIV or SHIV immunogens and viral challenge (Doria-Rose et al., 2003; Rosati et al., 2005; Yin et al., 2010), albeit there was significant improvement in immunogenicity with added cytokine genes (Barouch and Letvin, 2000) and with electroporation to enhance intramuscular uptake (Patel et al., 2010) and with improved vector design (Kulkarni et al., 2010). DNA vaccine immunogenicity is also greatly enhanced by protein boosting (Malherbe et al., 2011; Vaine et al., 2008). DNA vaccines have recently shown some promise for influenza vaccines in humans (Smith et al., 2010). They are attractive because they can deliver multiple antigens that can express genes in vivo to assemble noninfectious virus-like particles or native proteins that are difficult or impossible to produce by recombinant methodologies. They can be delivered multiple times without inducing anti-vector immunity that limits the effectiveness of viral vectors.

Another noteworthy area of active research in nonhuman primate models is the development of protein immunogens and improved adjuvants. Novel adjuvants are needed to stimulate both innate and adaptive responses, and with a better understanding of TLR binding it may be possible to direct responses more effectively, while increasing the magnitude of the response. In addition to whole virion approaches to preserve native structures (Frank et al., 2002; Johnson et al., 1992; Lifson et al., 2002; Warfield et al., 2007b; Willey et al., 2003), novel and native scaffold approaches are being modeled to present epitopes out of context on the surface of heterologous proteins (Guenaga et al., 2011; van Montfort et al., 2011; Zolla-Pazner et al., 2011), often in multimeric arrays that B cells prefer (De Berardinis et al., 1999).

HIV/SIV

The overwhelming body of literature on HIV vaccines in macaques has been summarized a number of times in several recent reviews (Haigwood, 2009) and chapters and thus the committee is encouraged to avail themselves of the detailed information summarized in a recent review (Lifson and Haigwood, in press). Many vaccines have focused on SIV only, while some have also included HIV Envelope for SHIV challenge. These studies have provided key observations about the immune responses elicited and how these have correlated with protection from infection. A critical example of this is the T cell-focused vaccines based on a non-replicating recombinant human adenovirus (Ad5) expressing SIV Gag, Pol, and Nef. A simiar SIV Gag vaccine had an effect on plasma viremia, reducing it 1 or 2 $\log_{10}$ using a less pathogenic X4 SHIV challenge virus (Shiver et al., 2002). When these SIV vaccines expressing Gag, Pol, and Nef were tested using the most pathogenic challenge (SIVmac239), there was no evidence of control of viremia post infection (Casimiro et al., 2010), the result found in the STEP clinical trial (Buchbinder et al., 2008). However, until there is a definitively positive human vaccine trial with correlates of immunity, we will not know for certain which of these macaque infection models and/or which immune responses are predictive of protection (Haigwood, 2009; Morgan et al., 2008). At this stage in vaccine development, the field awaits with great interest the ongoing immunological analysis of the RV144 trial, which showed modest, transient efficacy in a subset of the trial participants, those with lowest risk of exposure (Rerks-Ngarm et al., 2009). A few key lessons from this field are highlighted below:

Combination vaccines show more promise than single entities Combination vaccines—prime-boost—with multiple antigens that stimulate T-cell and B-cell responses generally have been found to be consistently more immunogenic and more effective in resisting challenge or controlling viremia than single approaches, but even these experiments had different outcomes depending upon the challenge virus. Early success with highly immunogenic yet risky vaccinia virus led to the testing of attenuated poxviruses such as Modified Vaccinia Ankara (MVA) and the avian poxvirus ALVAC. For SIV and SHIV, there are many examples of combination poxvirus vaccines that include one or more vectors with or without DNA or a recombinant protein, and these have shown different degrees of viral control upon challenge, depending upon the virus used for the challenge (Doria-Rose et al., 2003; Hel et al., 2002; Pal et al., 2006; Patterson et al., 2003). The recent RV144 trial was designed with an avian poxvirus prime, followed by a ALVAC plus Env protein boost. This vaccine has shown some modest and transient protection in humans (Rerks-Ngarm et al., 2009)—was it predicted by macaque experiments? Because no exact replica of the RV144 design was tested, a closely similar experiment based on SIV immunogens is in progress in macaques at this writing so that comparisons can be made.

New vectors to stimulate effector T cell memory   Current promising SIV vaccines based on rhCMV have demonstrated an impressive ability to control viremia to nearly undetectable levels in approximately half of the macaques challenged with SIVmac239 (Hansen et al., 2009, 2010, 2011). The mechanism of this vaccine is the persistence of the vector, which is apparently years at least, and the very strong induction of effector T cells against the SIV gene products. Further research is underway to understand the bimodal effect of the vaccine (all-or-none effects on virus loads) as well as to combine this approach with other vaccines that could stimulate central memory, such as adenovirus vectors, and antibodies, such as recombinant adjuvanted proteins. To move this approach into the clinic, it will require making HIV antigens in an attenuated hCMV vector, since hCMV is a high-risk infection for fetuses and immune-compromised humans.

Immunity to adapted viruses may provide clues for non-adapted infection   Because both HIV-1 and HIV-2 are derived from naturally occurring SIVs, understanding the immune responses in the host to which they are adapted could yield clues as to whether immunity in the non-adapted host contributes to pathogenesis (Silvestri, 2009). SIV not only replicates well in its natural host (sooty mangabeys, *Cercocebus atys*) (Sodora et al., 2009), but there is also evidence for loss of gut CD4 T cells; thus the idea has been proposed that these cells may not be the major source of plasma virus, even though they are an early target for destruction (Lay et al., 2009). These studies suggest that inflammation induces CCR5 expression as a co-factor for pathogenesis, as CD4$^+$ cells that are also CCR5$^+$ are greatly reduced in blood and tissue in the naturally infected hosts (Pandrea et al., 2007). Immune activation in pathogenic infection with SIV and HIV has been shown to be related to enhanced microbial translocation, which is related to CCR5 expression and can be partially controlled by antiretroviral treatment (Brenchley et al., 2006); translocation is reduced in the sooty mangabeys and is likely to be a similar story with the other highly adapted viruses in African green monkeys (Pandrea et al., 2006).

Improved challenge models to better mimic low-dose mucosal exposure
Studies in macaques have extended early findings in chimpanzee models for HIV-1 that passive transfer of antibodies (monoclonals or IgG from infected animals) can be delivered by intravenous, intramuscular, or subcutaneous routes to directly demonstrate a role for antibodies in blocking infection (reviewed in Lifson and Haigwood, in press). These studies have informed the magnitude and breadth of antibodies that may be necessary for a fully effective prophylactic vaccine, but much of this work was based on high-dose mucosal challenges, in order to achieve infection of all control animals after a limited number (1-2) of challenges. A very important advance to our understanding has been furthered by the use of low-dose challenge models that more closely resemble human transmission (Keele et al., 2008). These newer studies indicate that much lower doses than previously tested can be effective in preventing viral acquisition upon repeated low-dose challenge (Hessell et al., 2009a, 2009b). The use of repeated low-dose challenge in vaccine studies increases the expense and time needed to obtain answers but may provide more realistic assessment of the types and magnitude of immunity needed for more typical human sexual exposure to HIV-1.

Gene therapy as proof of principle for antibody-mediated protection
One of the most innovative uses of the SIV/macaque model was to directly prove that neutralizing antibodies expressed in vivo could prevent infection. This question was critically important to address because work prior to this point had suggested that only extraordinarily high levels of antibodies would be effective in preventing infection. Gene therapy and adenovirus associated virus (AAV) expressing an SIV neutralizing monoclonal sFv were utilized to prevent intravenous infection by SIV

(Johnson et al., 2009). Not all macaques were able to resist infection, and this appeared to correlate with the persistent expression of the antibody in vivo. This experiment could have failed completely, or could have caused major pathogenic consequences in vivo, and thus was an excellent use of the SIV macaque system to test proof of principle. If the problems can be overcome, this may see further development.

Tuberculosis

Tuberculosis (TB) is a major threat worldwide, particularly with ~2 billion people latently infected. Identifying immune mechanisms that control the initial infection and prevent reactivation remain critical goals (summarized in Lin and Flynn, 2010). Examples of the contributions of macaques to understanding these issues includes a demonstration of the role of T cells in disease control via a CD8 depletion study in BCG-vaccinated macaques that were infected with *M. tuberculosis* (Chen et al., 2009). Diedrich et al. (2010) used cynomolgus macaques with latent TB co-infected with SIVmac251 to develop the first animal model of reactivated TB in HIV-infected humans to better explore these factors. All latent animals developed reactivated TB following SIV infection, with a variable time to reactivation (up to 11 months post-SIV). Reactivation was independent of virus load but correlated with depletion of peripheral T cells during acute SIV infection (Diedrich et al., 2010). Further studies on SIV and TB co-infection indicate that events during acute HIV infection are likely to include distortions in proinflammatory and anti-inflammatory T cell responses within the granuloma that have significant effects on reactivation of latent TB. In this study, mycobacteria-specific multifunctional T cells were better correlates of Ag load (i.e., disease status) than of protection (Mattila et al., 2011).

Smallpox and Monkeypox

Smallpox and monkeypox are closely related orthopoxviruses that differ in their pathogenicity for humans. Although less infectious and the cause of less mortality, monkeypox is still a problem for humans when zoonotic transmissions take place. Due to the eradication of smallpox, vaccines can no longer be tested in humans for the prevention of the infection and thus nonhuman primate (macaque) studies are necessary for licensing. MVA has shown strong protective effects in *M. fasicularis* (Earl et al., 2004) and was shown to be more effective than the standard smallpox vaccine. Protection correlated with the more rapid immune response to MVA, potentially related to the higher dose of MVA that can be tolerated safely (Earl et al., 2008). Another group showed by depletion studies that B cells are essential for protection in this model (Edghill-Smith et al., 2005), and went on to develop a DNA/protein approach that shows promise of being a safe and effective vaccine (Heraud et al., 2006).

Yellow Fever Virus

Although there is a vaccine for yellow fever virus (YFV-17D) in use since 1945 that was originally developed in the macaque, it is not a fully efficacious vaccine, and severe adverse events have been reported that may be related in some cases to impaired innate responses (Pulendran et al., 2008). The vaccine elicits long-lived persistent T (Akondy et al., 2009) and B cell responses (Poland et al., 1981), and systems biology has been applied to determine correlates of protective immunity (Querec et al., 2009). To test for improved vaccines, there is currently a model for yellow fever using the YFV-Dakar strain of virus that has previously been characterized as viscerotropic and capable of being lethal in rhesus macaques (Monath et al., 1981). Following challenge with YFV-Dakar, unvaccinated animals demonstrated fever, lymphocytopenia, and fulminant viscerotropic disease with multi-organ failure, resulting in death within 4-6 days after infection. Histological analysis of the liver demonstrated widespread neutrophil infiltration, councilman bodies, and severe tissue necrosis, which correlated well with ALT (alanine aminotransferase). In contrast, animals that were vaccinated can show protection against lethal challenge and there is evidence that novel inactivated vaccines may protect against viremia, at least below detection (<50 genome copies/mL of serum) (M. Slifka, personal communication). Whether these new vaccines will offer improved safety profiles along with broad efficacy in humans remains to be determined.

Ebola Virus

The acute hemorrhagic filoviruses Ebola and Marburg cause infections with very high mortality rates in humans and nonhuman primates. Several approaches have been tested in macaques, with a primary focus on Ebola virus (EBOV). These have included DNA-prime-adenovirus boost with glycoprotein and nucleoprotein, either in a prolonged

(Sullivan et al., 2000) or shortened regimen (Sullivan et al., 2003). Although effective, preexisting adenovirus immunity may limit the utility of this approach. Other approaches tested have included live attenuated vaccines based on vesicular stomatitis virus (VSV) (Jones et al., 2005), as well as parainfluenzavirus (Bukreyev et al., 2010) and virus-like particles (Warfield et al., 2007a, 2007b). Discordant results of vaccine efficacy studies between mice and nonhuman primates have been observed with Ebola vaccines and have underscored the importance of examining this question. Although IgG titers are correlated with protection from EBOV challenge, passive antibody transfer was not fully effective in protection in macaques, demonstrating that protection is multifaceted. This subject is summarized in an excellent recent review that lays out the argument for using the "animal rule" for vaccine approval (Sullivan et al., 2009). Ultimately, the development of a vaccine for humans based on a replication defective Ad5 platform has shown significant promise and good immunogenicity (Ledgerwood et al., 2010).

CONCLUSIONS AND FUTURE USES OF NONHUMAN PRIMATE MODELS, INCLUDING THE CHIMPANZEE

- Chimpanzees have been essential for the study of human pathogens that do not infect lower species or reproduce key features of human disease. It is difficult to exclude the possibility that emerging infectious diseases will have a similar highly-restricted host range and thus be difficult to model in lower primates.
- There are as yet no vaccines for many of the human infectious diseases that have benefited from chimpanzee studies, including HCV, RSV, and malaria. All three remain important public health problems and there are high hurdles to successful vaccine development. These hurdles include a poor understanding of how to (1) vaccinate against highly mutable viruses that establish persistence (e.g., HCV), (2) safely balance vaccine immunogenicity with attenuation (e.g., RSV), and (3) select antigens for vaccination against a parasite with a complex life cycle and poorly understood strategies for immune evasion (e.g., malaria). It should be emphasized that even for an existing successful vaccine, unexpected adaptation of a virus like HBV can create vaccine escape variants. The chimpanzee model has in the recent

past, and may again in the future, be important to test the threat that such adaptations present to public health.

- Monoclonal antibodies and other biologicals designed to modulate inflammation and immunity in infectious and non-infectious diseases are routinely assessed for off-target effects in chimpanzees. The chimpanzee is valuable for these pre-clinical studies because unexpected cross-reactivity with orthologous proteins is more likely to be revealed. Humanized monoclonal antibodies are also less likely to elicit a neutralizing humoral response in chimpanzees when compared with more distant non-human primate species. This use of the animal model should accelerate as new targets for intervention are identified. Antibodies against T cell activating and inhibitory receptors that might modulate immune function in autoimmunity, cancer, and infectious diseases provide a prime example. For instance, combinations of inhibitory receptor blocking antibodies may be useful to restore immunity in chronic hepatitis B, but the animal model will be important to assess effectiveness and risk in a disease that is often subclinical and slowly progressive.

- The very close genetic relationship between humans and chimpanzees affords the opportunity to define the molecular pathogenesis of infections caused by viruses, microbes, and parasites that afflict humans. Genomic and proteomic technologies can be applied to understand host responses over the course of acute and (where relevant) persistent infection in primary target tissues. These tissues are often not available from humans because biopsies are not medically indicated. Also, critical aspects of innate and adaptive immune responses may be missed in humans because early stages of the infection are asymptomatic. Thus, chimpanzees provide a means to define immune responses at time points and locations that are inaccessible in humans.

- The close similarity of the macaque and human immune systems and the susceptibility of *Macaca species* to many human pathogens have together afforded opportunities to explore systematic comparison of multiple approaches for vaccine design, delivery, and comparative analyses of immunogenicity and responses to challenge. Furthermore, the relative availability of macaques for experimentation has allowed the discrimination of the contributions of different arms of the immune response by passive transfer of antibodies or transient depletion of specific subsets of cells.

- Having access to a broad range, or "full spectrum" of nonhuman primates has been enormously useful in order to understand how pathogens affect different species, in order to gain an understanding of the interplay between the host and the pathogen. Disease models differ in robustness depending on which species is used, so it is critical to have multiple species available. Understanding control in a species that has adapted to a virus may yield insights into novel therapeutics.

- "Failure" of an imperfect model can lead to a better understanding of the host-pathogen relationship. Mismatched outcomes between humans and animals, once understood, can provide insight into the pathogenesis of infection in humans. They can also lead to important refinements in a nonhuman primate model so that it better reflects the situation in humans. An example is high- versus low-dose mucosal challenge with SIV and SHIV in macaques. There was very strong resistance to changing challenge modalities based on the cost of requiring many more animals per group, time, manpower and virus stocks needed to deliver daily challenges for weeks or months, and statistical complications due to different times of acquisition. When it was demonstrated that typical sexual transmission of HIV generally results in a single or few founder viruses, then there was a stronger scientific rationale for the lower-dose challenges that are typically used today. This has led to encouraging news that protection from this type of challenge and indeed protection in humans may be more attainable, based on the amounts of antibodies needed for protection in macaques.

BIBLIOGRAPHY

Abe, K., G. Inchauspe, T. Shikata, and A. M. Prince. 1992. Three different patterns of hepatitis C virus infection in chimpanzees. *Hepatology* 15:690-695.

Abi-Rached, L., A. K. Moesta, R. Rajalingam, L. A. Guethlein, and P. Parham. 2010. Human-specific evolution and adaptation led to major qualitative differences in the variable receptors of human and chimpanzee natural killer cells. *PLoS Genet* 6:e1001192.

Aggarwal, R. 2011. Hepatitis E: Historical, contemporary and future perspectives. *J Gastroenterol Hepatol* 26(Suppl 1):72-82.

Akondy, R. S., N. D. Monson, J. D. Miller, S. Edupuganti, D. Teuwen, H. Wu, F. Quyyumi, S. Garg, J. D. Altman, C. Del Rio, H. L. Keyserling, A. Ploss, C. M. Rice, W. A. Orenstein, M. J. Mulligan, and R. Ahmed. 2009. The yellow fever virus vaccine induces a broad and polyfunctional human memory CD8[+] T cell response. *J Immunol* 183:7919-7930.

Alter, H. J., R. H. Purcell, P. V. Holland, and H. Popper. 1978. Transmissible agent in non-A, non-B hepatitis. *Lancet* 1:459-463.

Alter, H. J., J. W. Eichberg, H. Masur, W. C. Saxinger, R. Gallo, A. M. Macher, H. C. Lane, and A. S. Fauci. 1984. Transmission of HTLV-III infection from human plasma to chimpanzees: An animal model for AIDS. *Science* 226:549-552.

Amako, Y., K. Tsukiyama-Kohara, A. Katsume, Y. Hirata, S. Sekiguchi, Y. Tobita, Y. Hayashi, T. Hishima, N. Funata, H. Yonekawa, and M. Kohara. 2010. Pathogenesis of hepatitis C virus infection in Tupaia belangeri. *J Virol* 84:303-311.

Asabe, S., S. F. Wieland, P. K. Chattopadhyay, M. Roederer, R. E. Engle, R. H. Purcell, and F.V. Chisari. 2009. The size of the viral inoculum contributes to the outcome of hepatitis B virus infection. *J Virol* 83:9652-9662.

Aspinall, R., J. Pido-Lopez, N. Imami, S. M. Henson, P. T. Ngom, M. Morre, H. Niphuis, E. Remarque, B. Rosenwirth, and J. L. Heeney. 2007. Old rhesus macaques treated with interleukin-7 show increased TREC levels and respond well to influenza vaccination. *Rejuv Res* 10:5-17.

Baba, T. W., Y. S. Jeong, D. Penninck, R. Bronson, M. F. Greene, and R. M. Ruprecht. 1995. Pathogenicity of live, attenuated SIV after mucosal infection of neonatal macaques. *Science* 267:1820-1825.

Bach, F. H., M. A. Engstrom, M. L. Bach, and K. W. Sell. 1972. Histocompatibility matching. VII. Mixed leukocyte cultures between chimpanzee and man. *Transplant Proc* 4:97-100.

Balner, H., A. van Leeuwen, H. Dersjant, and J. J. van Rood. 1967. Defined leukocyte antigens of chimpanzees: Use of chimpanzee isoantisera for leukocyte typing in man. *Transplantation* 5:624-642.

Barber, D. L., E. J. Wherry, R. J. Greenwald, A. H. Sharpe, G. J. Freeman, and R. Ahmed. 2004. PD-L1 blockade restores CD8 T cell function during chronic viral infection [Abstract]. *Clinical and Investigative Medicine* 27:8C.

Barker, L. F., F. V. Chisari, P. P. McGrath, D. W. Dalgard, R. L. Kirschstein, J. D. Almeida, T. S. Edington, D. G. Sharp, and M. R. Peterson. 1973. Transmission of type B viral hepatitis to chimpanzees. *J Infect Dis* 127:648-662.

Barker, L. F., J. E. Maynard, R. H. Purcell, J. H. Hoofnagle, K. R. Berquist, and W. T. London. 1975a. Viral hepatitis, type B, in experimental animals. *Am J Med Sci* 270:189-195.

Barker, L. F., J. E. Maynard, R. H. Purcell, J. H. Hoofnagle, K. R. Berquist, W. T. London, R. J. Gerety, and D. H. Krushak. 1975b. Hepatitis B virus infection in chimpanzees: Titration of subtypes. *J Infect Dis* 132:451-458.

Barouch, D. H., and N. L. Letvin. 2000. Cytokine-induced augmentation of DNA vaccine-elicited SIV-specific immunity in rhesus monkeys. *Dev Biol (Basel)* 104:85-92.

Barr, C. S., S. A. Chen, M. L. Schwandt, S. G. Lindell, H. Sun, S. J. Suomi, and M. Heilig. 2010. Suppression of alcohol preference by naltrexone in the rhesus macaque: A critical role of genetic variation at the micro-opioid receptor gene locus. *Biol Psychiat* 67:78-80.

Barreiro, L. B., and L. Quintana-Murci. 2010. From evolutionary genetics to human immunology: How selection shapes host defence genes. *Nat Rev Genet* 11:17-30.

Barreiro, L. B., J. C. Marioni, R. Blekhman, M. Stephens, and Y. Gilad. 2010. Functional comparison of innate immune signaling pathways in primates. *PLoS Genet* 6:e1001249.

Barth, H., J. Rybczynska, R. Patient, Y. Choi, R. K. Sapp, T. F. Baumert, K. Krawczynski, and T. J. Liang. 2011. Both innate and adaptive immunity mediate protective immunity against hepatitis C virus infection in chimpanzees. *Hepatology* 54:1135-1148.

Bassett, S.E., K.M. Brasky, and R.E. Lanford. 1998. Analysis of hepatitis C virus-inoculated chimpanzees reveals unexpected clinical profiles. *J Virol* 72:2589-2599.

Ben-Yehudah, A., C. A. T. Easley, B. P. Hermann, C. Castro, C. Simerly, K. E. Orwig, S. Mitalipov, and G. Schatten. 2010. Systems biology discoveries using non-human primate pluripotent stem and germ cells: Novel gene and genomic imprinting interactions as well as unique expression patterns. *Stem Cell Research & Therapy* 1:24.

Berman, P., J. Groopman, T. Gregory, P. Clapham, R. Weiss, R. Ferriarri, L. Riddle, C. Shimasaki, C. Lucas, L. Lasky, and J. Eichberg. 1988. HIV-1 challenge of chimpanzees immunized with recombinant gp120. *P Natl Acad Sci USA* 85:5200.

Bethea, C. L., J. M. Streicher, K. Coleman, F. K. Pau, R. Moessner, and J. L. Cameron. 2004. Anxious behavior and fenfluramine-induced prolactin secretion in young rhesus macaques with different alleles of the serotonin reuptake transporter polymorphism (5HTTLPR). *Behav Genet* 34:295-307.

Bimber, B. N., A. J. Moreland, R. W. Wiseman, A. L. Hughes, and D. H. O'Connor. 2008. Complete characterization of killer Ig-like receptor (KIR) haplotypes in Mauritian cynomolgus macaques: Novel insights into nonhuman primate KIR gene content and organization. *J Immunol* 181:6301-6308.

Blount, R. E., Jr., J. A. Morris, and R. E. Savage. 1956. Recovery of cytopathogenic agent from chimpanzees with coryza. *Proc Soc Exp Biol Med* 92:544-549.

Boni, C., P. Fisicaro, C. Valdatta, B. Amadei, P. Di Vincenzo, T. Giuberti, D. Laccabue, A. Zerbini, A. Cavalli, G. Missale, A. Bertoletti, and C. Ferrari. 2007. Characterization of hepatitis B virus (HBV)-specific T-cell dysfunction in chronic HBV infection. *J Virol* 81:4215-4225.

Bontrop, R. E., and D. I. Watkins. 2005. MHC polymorphism: AIDS susceptibility in non-human primates. *Trends Immunol* 26:227-233.

Bowen, D. G., and C. M. Walker. 2005. Adaptive immune responses in acute and chronic hepatitis C virus infection. *Nature* 436:946-952.

Boyson, J. E., C. Shufflebotham, L. F. Cadavid, J. A. Urvater, L. A. Knapp, A. L. Hughes, and D. I. Watkins. 1996. The MHC class I genes of the rhesus monkey. Different evolutionary histories of MHC class I and II genes in primates. *J Immunol* 156:4656-4665.

Bradley, D. W., J. E. Maynard, H. Popper, E. H. Cook, J. W. Ebert, K. A. McCaustland, C. A. Schable, and H. A. Fields. 1983. Posttransfusion non-A, non-B hepatitis: Physicochemical properties of two distinct agents. *J Infect Dis* 148:254-265.

Bradley, D. W., K. A. McCaustland, E. H. Cook, C. A. Schable, J. W. Ebert, and J. E. Maynard. 1985. Posttransfusion non-A, non-B hepatitis in chimpanzees. Physicochemical evidence that the tubule-forming agent is a small, enveloped virus. *Gastroenterology* 88:773-779.

Brenchley, J. M., D. A. Price, T. W. Schacker, T. E. Asher, G. Silvestri, S. Rao, Z. Kazzaz, E. Bornstein, O. Lambotte, D. Altmann, B. R. Blazar, B. Rodriguez, L. Teixeira-Johnson, A. Landay, J. N. Martin, F. M. Hecht, L. J. Picker, M. M. Lederman, S. G. Deeks, and D. C. Douek. 2006. Microbial translocation is a cause of systemic immune activation in chronic HIV infection. *Nat Med* 12:1365-1371.

Buchbinder, S. P., D. V. Mehrotra, A. Duerr, D. W. Fitzgerald, R. Mogg, D. Li, P. B. Gilbert, J. R. Lama, M. Marmor, C. Del Rio, M. J. McElrath, D. R. Casimiro, K. M. Gottesdiener, J. A. Chodakewitz, L. Corey, and M. N. Robertson. 2008. Efficacy assessment of a cell-mediated immunity HIV-1 vaccine (the STEP Study): A double-blind, randomised, placebo-controlled, test-of-concept trial. *Lancet* 372:1881-1893.

Bukh, J., C. L. Apgar, S. Govindarajan, S. U. Emerson, and R. H. Purcell. 2001. Failure to infect rhesus monkeys with hepatitis C virus strains of genotypes 1a, 2a or 3a. *J Viral Hepat* 8:228-231.

Bukreyev, A. A., J. M. Dinapoli, L. Yang, B. R. Murphy, and P. L. Collins. 2010. Mucosal parainfluenza virus-vectored vaccine against Ebola virus replicates in the respiratory tract of vector-immune monkeys and is immunogenic. *Virology* 399:290-298.

Buynak, E. B., R. R. Roehm, A. A. Tytell, A. U. Bertland, 2nd, G. P. Lampson, and M. R. Hilleman. 1976a. Development and chimpanzee testing of a vaccine against human hepatitis B. *Proc Soc Exp Biol Med* 151:694-700.

Buynak, E. B., R. R. Roehm, A. A. Tytell, A. U. Bertland, 2nd, G. P. Lampson, and M. R. Hilleman. 1976b. Vaccine against human hepatitis B. *JAMA* 235:2832-2834.

Callendret, B., and C. Walker. 2011. A siege of hepatitis: Immune boost for viral hepatitis. *Nat Med* 17:252-253.

Casimiro, D. R., K. Cox, A. Tang, K. J. Sykes, M. Feng, F. Wang, A. Bett, W. A. Schleif, X. Liang, J. Flynn, T. W. Tobery, K. Wilson, A. Finnefrock, L. Huang, S. Vitelli, J. Lin, D. Patel, M. E. Davies, G. J. Heidecker, D. C. Freed, S. Dubey, D. H. O'Connor, D. I. Watkins, Z. Q. Zhang, and J. W. Shiver. 2010. Efficacy of multivalent adenovirus-based vaccine against simian immunodeficiency virus challenge. *J Virol* 84:2996-3003.

Chanock, R. M., and L. Finberg. 1957. Recovery from infants with respiratory illness of a virus related to chimpanzee coryza agent (CCA). II. Epidemiological aspects of infection in infants and young children. *Am J Hyg* 66:291-300.

Chanock, R. M., B. Roizman, and R. Myers. 1957. Recovery from infants with respiratory illness of a virus related to chimpanzee coryza agent. I. Isolation, properties and characterization. *Am J Hyg* 66:281-290.

Chen, C. Y., D. Huang, R. C. Wang, L. Shen, G. Zeng, S. Yao, Y. Shen, L. Halliday, J. Fortman, M. McAllister, J. Estep, R. Hunt, D. Vasconcelos, G. Du, S. A. Porcelli, M. H. Larsen, W. R. Jacobs, Jr., B. F. Haynes, N. L. Letvin, and Z. W. Chen. 2009. A critical role for CD8 T cells in a nonhuman primate model of tuberculosis. *PLoS Pathog* 5:e1000392.

Choo, Q. L., G. Kuo, A. J. Weiner, L. R. Overby, D. W. Bradley, and M. Houghton. 1989. Isolation of a cDNA clone derived from a blood-borne non-A, non-B viral hepatitis genome. *Science* 244:359-362.

Cohen, J. 1999. AIDS vaccine. Chimps and lethal strain a bad mix. *Science* 286:1454-1455.

Cole, K. S., J. L. Rowles, B. A. Jagesrski, M. Murphey-Corb, T. Unangst, J. E. Clements, J. Robinson, M. S. Wyand, R. C. Desrosiers, and R. C. Montelaro. 1997. Evolution of envelope-specific antibody responses in monkeys experimentally infected or immunized with simian immunodeficiency virus and its association with the development of protective immunity. *J Virol* 71:5069-5079.

Cole, K. S., M. Alvarez, D. H. Elliott, H. Lam, E. Martin, T. Chau, K. Micken, J. L. Rowles, J. E. Clements, M. Murphey-Corb, R. C. Montelaro, and J. E. Robinson. 2001. Characterization of neutralization epitopes of simian immunodeficiency virus (SIV) recognized by rhesus monoclonal antibodies derived from monkeys infected with an attenuated SIV strain. *Virology* 290:59-73.

Conley, A. J., J. A. Kessler, II, L. J. Boots, P. M. McKenna, W. A. Schleif, E. A. Emini, G. E. Mark, III, H. Katinger, E. K. Cobb, S. M. Lunceford, S. R. Rouse, and K. K. Murthy. 1996. The consequence of passive administration of an anti-human immunodeficiency virus type 1 neutralizing monoclonal antibody before challenge of chimpanzees with a primary virus isolate. *J Virol* 70:6751-6758.

Cooper, S., H. Kowalski, A. L. Erickson, K. Arnett, A. M. Little, C. M. Walker, and P. Parham. 1996. The presentation of a hepatitis C viral peptide by distinct major histocompatibility complex class I allotypes from two chimpanzee species. *J Exp Med* 183:663-668.

Crowe, J. E., Jr., P. L. Collins, W. T. London, R. M. Chanock, and B. R. Murphy. 1993. A comparison in chimpanzees of the immunogenicity and efficacy of live attenuated respiratory syncytial virus (RSV) temperature-sensitive mutant vaccines and vaccinia virus recombinants that express the surface glycoproteins of RSV. *Vaccine* 11:1395-1404.

Crowe, J. E., Jr., P. T. Bui, W. T. London, A. R. Davis, P. P. Hung, R. M. Chanock, and B. R. Murphy. 1994. Satisfactorily attenuated and protective mutants derived from a partially attenuated cold-passaged respiratory syncytial virus mutant by introduction of additional attenuating mutations during chemical mutagenesis. *Vaccine* 12:691-699.

Currier, J. R., K. S. Stevenson, P. J. Kehn, K. Zheng, V. M. Hirsch, and M. A. Robinson. 1999. Contributions of $CD4^+$, $CD8^+$, and $CD4^+CD8^+$ T cells to skewing within the peripheral T cell receptor beta chain repertoire of healthy macaques. *Hum Immunol* 60:209-222.

Dahari, H., S. M. Feinstone, and M. E. Major. 2010. Meta-analysis of hepatitis C virus vaccine efficacy in chimpanzees indicates an importance for structural proteins. *Gastroenterology* 139:965-974.

Daniel, M. D., N. L. Letvin, N. W. King, M. Kannagi, P. K. Sehgal, R. D. Hunt, P. J. Kanki, M. Essex, and R. C. Desrosiers. 1985. Isolation of T-cell tropic HTLV-III-like retrovirus from macaques. *Science* 228:1201-1204.

Daubersies, P., A. W. Thomas, P. Millet, K. Brahimi, J. A. Langermans, B. Ollomo, L. BenMohamed, B. Slierendregt, W. Eling, A. Van Belkum, G. Dubreuil, J. F. Meis, C. Guerin-Marchand, S. Cayphas, J. Cohen, H. Gras-Masse, and P. Druilhe. 2000. Protection against Plasmodium falciparum malaria in chimpanzees by immunization with the conserved pre-erythrocytic liver-stage antigen 3. *Nat Med* 6:1258-1263.

Daubersies, P., B. Ollomo, J. P. Sauzet, K. Brahimi, B. L. Perlaza, W. Eling, H. Moukana, P. Rouquet, C. de Taisne, and P. Druilhe. 2008. Genetic immunisation by liver stage antigen 3 protects chimpanzees against malaria despite low immune responses. *PloS one* 3:e2659.

De Berardinis, P., L. D'Apice, A. Prisco, M. N. Ombra, P. Barba, G. Del Pozzo, S. Petukhov, P. Malik, R. N. Perham, and J. Guardiola. 1999. Recognition of HIV-derived B and T cell epitopes displayed on filamentous phages. *Vaccine* 17:1434-1441.

de Groot, N. G., C. M. Heijmans, N. de Groot, G. G. Doxiadis, N. Otting, and R. E. Bontrop. 2009. The chimpanzee Mhc-DRB region revisited: Gene content, polymorphism, pseudogenes, and transcripts. *Mol Immunol* 47:381-389.

de Groot, N. G., C. M. Heijmans, Y. M. Zoet, A. H. de Ru, F. A. Verreck, P. A. van Veelen, J. W. Drijfhout, G. G. Doxiadis, E. J. Remarque, Doxiadis, II, J. J. van Rood, F. Koning, and R. E. Bontrop. 2010. AIDS-protective HLA-B*27/B*57 and chimpanzee MHC class I molecules target analogous conserved areas of HIV-1/SIVcpz. *P Natl Acad Sci USA* 107:15175-15180.

Diedrich, C. R., J. T. Mattila, E. Klein, C. Janssen, J. Phuah, T. J. Sturgeon, R. C. Montelaro, P. L. Lin, and J. L. Flynn. 2010. Reactivation of latent tuberculosis in cynomolgus macaques infected with SIV is associated with early peripheral T cell depletion and not virus load. *PloS one* 5:e9611.

Dienstag, J. L., S. M. Feinstone, R. H. Purcell, J. H. Hoofnagle, L. F. Barker, W. T. London, H. Popper, J. M. Peterson, and A. Z. Kapikian. 1975. Experimental infection of chimpanzees with hepatitis A virus. *J Infect Dis* 132:532-545.

Dienstag, J. L., H. Popper, and R. H. Purcell. 1976. The pathology of viral hepatitis types A and B in chimpanzees. A comparison. *Am J Patho* 85:131-148.

Donahue, R. E., and C. E. Dunbar. 2001. Update on the use of nonhuman primate models for preclinical testing of gene therapy approaches targeting hematopoietic cells. *Hum Gene Ther* 12:607-617.

Doria-Rose, N. A., C. Ohlen, P. Polacino, C. C. Pierce, M. T. Hensel, L. Kuller, T. Mulvania, D. Anderson, P. D. Greenberg, S. L. Hu, and N. L. Haigwood. 2003. Multigene DNA priming-boosting vaccines protect macaques from acute CD4+-T-cell depletion after simian-human immunodeficiency virus SHIV89.6P mucosal challenge. *J Virol* 77:11563-11577.

Doria-Rose, N. A., R. M. Klein, M. M. Manion, S. O'Dell, A. Phogat, B. Chakrabarti, C. W. Hallahan, S. A. Migueles, J. Wrammert, R. Ahmed, M. Nason, R. T. Wyatt, J. R. Mascola, and M. Connors. 2009. Frequency and phenotype of human immunodeficiency virus envelope-specific B cells from patients with broadly cross-neutralizing antibodies. *J Virol* 83:188-199.

Doxiadis, G. G., C. M. Heijmans, N. Otting, and R. E. Bontrop. 2007. MIC gene polymorphism and haplotype diversity in rhesus macaques. *Tissue Antigens* 69:212-219.

Earl, P. L., J. L. Americo, L. S. Wyatt, L. A. Eller, J. C. Whitbeck, G. H. Cohen, R. J. Eisenberg, C. J. Hartmann, D. L. Jackson, D. A. Kulesh, M. J. Martinez, D. M. Miller, E. M. Mucker, J. D. Shamblin, S. H. Zwiers, J. W.

Huggins, P. B. Jahrling, and B. Moss. 2004. Immunogenicity of a highly attenuated MVA smallpox vaccine and protection against monkeypox. *Nature* 428:182-185.

Earl, P. L., J. L. Americo, L. S. Wyatt, O. Espenshade, J. Bassler, K. Gong, S. Lin, E. Peters, L. Rhodes, Jr., Y. E. Spano, P. M. Silvera, and B. Moss. 2008. Rapid protection in a monkeypox model by a single injection of a replication-deficient vaccinia virus. *P Natl Acad Sci USA* 105:10889-10894.

Edghill-Smith, Y., H. Golding, J. Manischewitz, L. R. King, D. Scott, M. Bray, A. Nalca, J. W. Hooper, C. A. Whitehouse, J. E. Schmitz, K. A. Reimann, and G. Franchini. 2005. Smallpox vaccine-induced antibodies are necessary and sufficient for protection against monkeypox virus. *Nature Medicine* 11:740-747.

Eichberg, J. W., J. M. Zarling, H. J. Alter, J. A. Levy, P. W. Berman, T. Gregory, L. A. Lasky, J. McClure, K. E. Cobb, P. A. Moran, S. Hu, R. C. Kennedy, T. C. Chanh, and G. R. Dreesman. 1987. T-cell responses to human immunodeficiency virus (HIV) and its recombinant antigens in HIV-infected chimpanzees. *J Virol* 61:3804-3808.

El-Amad, Z., K. K. Murthy, K. Higgins, E. K. Cobb, N. L. Haigwood, J. A. Levy, and K. S. Steimer. 1995. Resistance of chimpanzees immunized with recombinant gp120SF2 to challenge by HIV-1SF2. *AIDS* 9:1313-1322.

Emini, E. A., P. L. Nara, W. A. Schlief, J. A. Lewis, J. P. Davide, D. R. Lee, J. Kessler, S. Conley, S. Matsushita, S. D. Putney, R. J. Gerety, and J. W. Eichberg. 1990. Antibody-mediated *in vitro* neutralization of human immunodeficiency virus type 1 abolishes infectivity for chimpanzees. *J Virol* 64:3674-3678.

Estep, R. D., I. Messaoudi, M. A. O'Connor, H. Li, J. Sprague, A. Barron, F. Engelmann, B. Yen, M. F. Powers, J. M. Jones, B. A. Robinson, B. U. Orzechowska, M. Manoharan, A. Legasse, S. Planer, J. Wilk, M. K. Axthelm, and S. W. Wong. 2011. Deletion of the monkeypox inhibitor of complement enzymes locus impacts the adaptive immune response to monkeypox virus in a non human model of infection. *J Virol*.

Fang, X., Y. Zhang, R. Zhang, L. Yang, M. Li, K. Ye, X. Guo, J. Wang, and B. Su. 2011. Genome sequence and global sequence variation map with 5.5 million SNPs in Chinese rhesus macaque. *Genome Biol* 12:R63.

Farci, P., H. J. Alter, S. Govindarajan, D. C. Wong, R. Engle, R. R. Lesniewski, I. K. Mushahwar, S. M. Desai, R. H. Miller, N. Ogata, and et al. 1992. Lack of protective immunity against reinfection with hepatitis C virus. *Science* 258:135-140.

Feinstone, S. M., A. Z. Kapikian, R. H. Purcell, H. J. Alter, and P. V. Holland. 1975. Transfusion-associated hepatitis not due to viral hepatitis type A or B. *N Engl J Med* 292:767-770.

Feinstone, S. M., R. J. Daemer, I. D. Gust, and R. H. Purcell. 1983. Live attenuated vaccine for hepatitis A. *Dev Biol Stand* 54:429-432.

Folgori, A., S. Capone, L. Ruggeri, A. Meola, E. Sporeno, B. B. Ercole, M. Pezzanera, R. Tafi, M. Arcuri, E. Fattori, A. Lahm, A. Luzzago, A. Vitelli, S. Colloca, R. Cortese, and A. Nicosia. 2006. A T-cell HCV vaccine eliciting effective immunity against heterologous virus challenge in chimpanzees. *Nat Med* 12:190-197.

Frank, I., M. Piatak, Jr., H. Stoessel, N. Romani, D. Bonnyay, J. D. Lifson, and M. Pope. 2002. Infectious and whole inactivated simian immunodeficiency viruses interact similarly with primate dendritic cells (DCs): differential intracellular fate of virions in mature and immature DCs. *J Virol* 76:2936-2951.

Fultz, P. N., H. M. McClure, H. Daugharty, A. Brodie, C. R. McGrath, B. Swenson, and D. P. Francis. 1986. Vaginal transmission of human immunodeficiency virus (HIV) to a chimpanzee. *J Infect Dis* 154:896-900.

Fultz, P. N., A. Srinivasan, C. R. Greene, D. Butler, R. B. Swenson, and H. M. McClure. 1987. Superinfection of a chimpanzee with a second strain of human immunodeficiency virus. *J Virol* 61:4026-4029.

Gao, F., D. L. Robertson, C. D. Carruthers, Y. Li, E. Bailes, L. G. Kostrikis, M. O. Salminen, F. Bibollet-Ruche, M. Peeters, D. D. Ho, G. M. Shaw, P. M. Sharp, and B. H. Hahn. 1999. Origin of HIV-1 in the chimpanzee Pan troglodytes troglodytes. *Nature* 397:436-441.

Gao, X., Y. Jiao, L. Wang, X. Liu, W. Sun, B. Cui, Z. Chen, and Y. Zhao. 2010. Inhibitory KIR and specific HLA-C gene combinations confer susceptibility to or protection against chronic hepatitis B. *Clin Immunol* 137:139-146.

Genesca, M., M. B. McChesney, and C. J. Miller. 2009. Antiviral CD8[+] T cells in the genital tract control viral replication and delay progression to AIDS after vaginal SIV challenge in rhesus macaques immunized with virulence attenuated SHIV 89.6. *J Intern Med* 265:67-77.

Gibbs, R. A., J. Rogers, M. G. Katze, R. Bumgarner, G. M. Weinstock, E. R. Mardis, K. A. Remington, R. L. Strausberg, J. C. Venter, R. K. Wilson, M. A. Batzer, C. D. Bustamante, E. E. Eichler, M. W. Hahn, R. C. Hardison, K. D. Makova, W. Miller, A. Milosavljevic, R. E. Palermo, A. Siepel, J. M. Sikela, T. Attaway, S. Bell, K. E. Bernard, C. J. Buhay, M. N. Chandrabose, M. Dao, C. Davis, K. D. Delehaunty, Y. Ding, H. H. Dinh, S. Dugan-Rocha, L. A. Fulton, R. A. Gabisi, T. T. Garner, J. Godfrey, A. C. Hawes, J. Hernandez, S. Hines, M. Holder, J. Hume, S. N. Jhangiani, V. Joshi, Z. M. Khan, E. F. Kirkness, A. Cree, R. G. Fowler, S. Lee, L. R. Lewis, Z. Li, Y. S. Liu, S. M. Moore, D. Muzny, L. V. Nazareth, D. N. Ngo, G. O. Okwuonu, G. Pai, D. Parker, H. A. Paul, C. Pfannkoch, C. S. Pohl, Y. H. Rogers, S. J. Ruiz, A. Sabo, J. Santibanez, B. W. Schneider, S. M. Smith, E. Sodergren, A. F. Svatek, T. R. Utterback, S. Vattathil, W. Warren, C. S. White, A. T. Chinwalla, Y. Feng, A. L. Halpern, L. W. Hillier, X. Huang, P. Minx, J. O. Nelson, K. H. Pepin, X. Qin, G. G. Sutton, E. Venter, B. P. Walenz, J. W. Wallis, K. C. Worley, S. P. Yang, S. M. Jones, M. A. Marra, M. Rocchi, J. E. Schein, R. Baertsch, L. Clarke, M. Csuros, J. Glasscock, R.

A. Harris, P. Havlak, A. R. Jackson, H. Jiang, Y. Liu, D. N. Messina, Y. Shen, H. X. Song, T. Wylie, L. Zhang, E. Birney, K. Han, M. K. Konkel, J. Lee, A. F. Smit, B. Ullmer, H. Wang, J. Xing, R. Burhans, Z. Cheng, J. E. Karro, J. Ma, B. Raney, X. She, M. J. Cox, J. P. Demuth, L. J. Dumas, S. G. Han, J. Hopkins, A. Karimpour-Fard, Y. H. Kim, J. R. Pollack, T. Vinar, C. Addo-Quaye, J. Degenhardt, A. Denby, M. J. Hubisz, A. Indap, C. Kosiol, B. T. Lahn, H. A. Lawson, A. Marklein, R. Nielsen, E. J. Vallender, A. G. Clark, B. Ferguson, R. D. Hernandez, K. Hirani, H. Kehrer-Sawatzki, J. Kolb, S. Patil, L. L. Pu, Y. Ren, D. G. Smith, D. A. Wheeler, I. Schenck, E. V. Ball, R. Chen, D. N. Cooper, B. Giardine, F. Hsu, W. J. Kent, A. Lesk, D. L. Nelson, E. O'Brien W, K. Prufer, P. D. Stenson, J. C. Wallace, H. Ke, X. M. Liu, P. Wang, A. P. Xiang, F. Yang, G. P. Barber, D. Haussler, D. Karolchik, A. D. Kern, R. M. Kuhn, K. E. Smith, and A. S. Zwieg. 2007. Evolutionary and biomedical insights from the rhesus macaque genome. *Science* 316:222-234.

Giles, G. R., H. J. Boehmig, H. Amemiya, C. G. Halgrimson, and T. E. Starzl. 1970. Clinical heterotransplantation of the liver. *Transplant Proc* 2:506-512.

Glamann, J., D. R. Burton, P. W. Parren, H. J. Ditzel, K. A. Kent, C. Arnold, D. Montefiori, and V. M. Hirsch. 1998. Simian immunodeficiency virus (SIV) envelope-specific Fabs with high-level homologous neutralizing activity: recovery from a long-term-nonprogressor SIV-infected macaque. *J Virol* 72:585-592.

Golden-Mason, L., B. Palmer, J. Klarquist, J. A. Mengshol, N. Castelblanco, and H. R. Rosen. 2007. Upregulation of PD-1 expression on circulating and intrahepatic HCV-specific CD8[+] T cells associated with reversible immune dysfunction. *J Virol* 81:9249-9258.

Good, M. F. 2011. Our impasse in developing a malaria vaccine. *Cell Mol Life Sci* 68:1105-1113.

Good, M. F., and D. L. Doolan. 2010. Malaria vaccine design: Immunological considerations. *Immunity* 33:555-566.

Goulder, P. J., and D. I. Watkins. 2008. Impact of MHC class I diversity on immune control of immunodeficiency virus replication. *Nat Rev Immunol* 8:619-630.

Graham, B. S. 2011. Biological challenges and technological opportunities for respiratory syncytial virus vaccine development. *Immunol Rev* 239:149-166.

Grakoui, A., N. H. Shoukry, D. J. Woollard, J. H. Han, H. L. Hanson, J. Ghrayeb, K. K. Murthy, C. M. Rice, and C. M. Walker. 2003. HCV persistence and immune evasion in the absence of memory T cell help. *Science* 302:659-662.

Grove, K. L., B. E. Grayson, M. M. Glavas, X. Q. Xiao, and M. S. Smith. 2005. Development of metabolic systems. *Physiol Behav* 86:646-660.

Grunbaum, A. S. 1904. Report LXXXIII: Some Experiments on Enterica, Scarlet Fever, and Measles in the Chimpanzee: [A Preliminary Communication]. *Br Med J* 1:817-819.

Guenaga, J., P. Dosenovic, G. Ofek, D. Baker, W. R. Schief, P. D. Kwong, G. B. Karlsson Hedestam, and R. T. Wyatt. 2011. Heterologous epitope-scaffold prime:boosting immuno-focuses B cell responses to the HIV-1 gp41 2F5 neutralization determinant. *PLoS One* 6:e16074.

Guidotti, L. G., R. Rochford, J. Chung, M. Shapiro, R. Purcell, and F. V. Chisari. 1999. Viral clearance without destruction of infected cells during acute HBV infection. *Science* 284:825-829.

Ha, S. J., S. N. Mueller, E. J. Wherry, D. L. Barber, R. D. Aubert, A. H. Sharpe, G. J. Freeman, and R. Ahmed. 2008. Enhancing therapeutic vaccination by blocking PD-1-mediated inhibitory signals during chronic infection. *J Exp Med* 205:543-555.

Hahn, B. H., G. M. Shaw, K. M. De Cock, and P. M. Sharp. 2000. AIDS as a zoonosis: scientific and public health implications. *Science* 287:607-614.

Haigwood, N. L. 2009. Update on animal models for HIV research. *Eur J Immunol* 39:1994-1999.

Haining, W. N., and E. J. Wherry. 2010. Integrating genomic signatures for immunologic discovery. *Immunity* 32:152-161.

Hall, C. B., G. A. Weinberg, M. K. Iwane, A. K. Blumkin, K. M. Edwards, M. A. Staat, P. Auinger, M. R. Griffin, K. A. Poehling, D. Erdman, C. G. Grijalva, Y. Zhu, and P. Szilagyi. 2009. The burden of respiratory syncytial virus infection in young children. *N Engl J Med* 360:588-598.

Hansen, S. G., C. Vieville, N. Whizin, L. Coyne-Johnson, D. C. Siess, D. D. Drummond, A. W. Legasse, M. K. Axthelm, K. Oswald, C. M. Trubey, M. Piatak, Jr., J.D. Lifson, J. A. Nelson, M. A. Jarvis, and L. J. Picker. 2009. Effector memory T cell responses are associated with protection of rhesus monkeys from mucosal simian immunodeficiency virus challenge. *Nat Med* 15:293-299.

Hansen, S. G., C. J. Powers, R. Richards, A. B. Ventura, J. C. Ford, D. Siess, M. K. Axthelm, J. A. Nelson, M. A. Jarvis, L. J. Picker, and K. Fruh. 2010. Evasion of CD8$^+$ T cells is critical for superinfection by cytomegalovirus. *Science* 328:102-106.

Hansen, S. G., J. C. Ford, M. S. Lewis, A. B. Ventura, C. M. Hughes, L. Coyne-Johnson, N. Whizin, K. Oswald, R. Shoemaker, T. Swanson, A. W. Legasse, M. J. Chiuchiolo, C. L. Parks, M. K. Axthelm, J. A. Nelson, M. A. Jarvis, M. Piatak, Jr., J. D. Lifson, and L. J. Picker. 2011. Profound early control of highly pathogenic SIV by an effector memory T-cell vaccine. *Nature* 473:523-527.

Harouse, J. M., A. Gettie, T. Eshetu, R. C. Tan, R. Bohm, J. Blanchard, G. Baskin, and C. Cheng-Mayer. 2001. Mucosal transmission and induction of simian AIDS by CCR5-specific simian/human immunodeficiency virus SHIV(SF162P3). *J Virol* 75:1990-1995.

Hayton, K., D. Gaur, A. Liu, J. Takahashi, B. Henschen, S. Singh, L. Lambert, T. Furuya, R. Bouttenot, M. Doll, F. Nawaz, J. Mu, L. Jiang, L. H. Miller, and T. E. Wellems. 2008. Erythrocyte binding protein PfRH5 polymorphisms determine species-specific pathways of Plasmodium falciparum invasion. *Cell Host Microbe* 4:40-51.

Hel, Z., J. Nacsa, W. P. Tsai, A. Thornton, L. Giuliani, J. Tartaglia, and G. Franchini. 2002. Equivalent immunogenicity of the highly attenuated poxvirus-based ALVAC-SIV and NYVAC-SIV vaccine candidates in SIVmac251-infected macaques. *Virology* 304:125-134.

Heraud, J. M., Y. Edghill-Smith, V. Ayala, I. Kalisz, J. Parrino, V. S. Kalyanaraman, J. Manischewitz, L.R. King, A. Hryniewicz, C. J. Trindade, M. Hassett, W. P. Tsai, D. Venzon, A. Nalca, M. Vaccari, P. Silvera, M. Bray, B. S. Graham, H. Golding, J. W. Hooper, and G. Franchini. 2006. Subunit recombinant vaccine protects against monkeypox. *J Immunol* 177:2552-2564.

Hessell, A. J., P. Poignard, M. Hunter, L. Hangartner, D. M. Tehrani, W. K. Bleeker, P. W. Parren, P. A. Marx, and D. R. Burton. 2009a. Effective, low-titer antibody protection against low-dose repeated mucosal SHIV challenge in macaques. *Nat Med* 15:951-954.

Hessell, A. J., E. G. Rakasz, P. Poignard, L. Hangartner, G. Landucci, D. N. Forthal, W. C. Koff, D. I. Watkins, and D. R. Burton. 2009b. Broadly neutralizing human anti-HIV antibody 2G12 is effective in protection against mucosal SHIV challenge even at low serum neutralizing titers. *PLoS Pathog* 5:e1000433.

Hirsch, V. M., and J. D. Lifson. 2000. Simian immunodeficiency virus infection of monkeys as a model system for the study of AIDS pathogenesis, treatment, and prevention. *Adv Pharmacol* 49:437-477.

Hirsch, V. M., G. Dapolito, A. Hahn, J. Lifson, D. Montefiori, C. R. Brown, and R. Goeken. 1998. Viral genetic evolution in macaques infected with molecularly cloned simian immunodeficiency virus correlates with the extent of persistent viremia. *J Virol* 72:6482-6489.

Hoffman, S. L., and D. L. Doolan. 2000. Malaria vaccines-targeting infected hepatocytes. *Nat Med* 6:1218-1219.

Hu, X., H. S. Margolis, R. H. Purcell, J. Ebert, and B. H. Robertson. 2000. Identification of hepatitis B virus indigenous to chimpanzees. *P Natl Acad Sci USA* 97:1661-1664.

Huang, K. H., D. Bonsall, A. Katzourakis, E. C. Thomson, S. J. Fidler, J. Main, D. Muir, J. N. Weber, A. J. Frater, R. E. Phillips, O. G. Pybus, P. J. Goulder, M. O. McClure, G. S. Cooke, and P. Klenerman. 2010. B-cell depletion reveals a role for antibodies in the control of chronic HIV-1 infection. *Nature Commu* 1:102.

Jaeger, E. E., R. E. Bontrop, and J. S. Lanchbury. 1993. Nucleotide sequences, polymorphism and gene deletion of T cell receptor beta-chain constant regions of Pan troglodytes and Macaca mulatta. *J Immunol* 151:5301-5309.

Jaeger, E. E., R. E. Bontrop, P. Parham, E. J. Wickings, M. Kadwell, and J. S. Lanchbury. 1998. Characterization of chimpanzee TCRV gene polymorphism: How old are human TCRV alleles? *Immunogenetics* 47:115-123.

Jamil, K. M., and S. I. Khakoo. 2011. KIR/HLA interactions and pathogen immunity. *J Biomed Biotechnol* 2011:298348.

Jayaraman, P., T. Zhu, L. Misher, D. Mohan, L. Kuller, P. Polacino, B. A. Richardson, H. Bielefeldt-Ohmann, D. Anderson, S. L. Hu, and N. L. Haigwood. 2007. Evidence for persistent, occult infection in neonatal macaques following perinatal transmission of simian-human immunodeficiency virus SF162P3. *J Virol* 81:822-834.

Johnson, P. R., D. C. Montefiori, S. Goldstein, T. E. Hamm, J. Zhou, S. Kitov, N. L. Haigwood, L. Misher, W. T. London, J. L. Gerin, A. Allison, R. H. Purcell, R. M. Chanock, and V. M. Hirsch. 1992. Inactivated whole-virus vaccine derived from a proviral DNA clone of simian immunodeficiency virus induces high levels of neutralizing antibodies and confers protection against heterologous challenge. *P Natl Acad Sci USA* 89:2175-2179.

Johnson, P. R., B. C. Schnepp, J. Zhang, M. J. Connell, S. M. Greene, E. Yuste, R. C. Desrosiers, and K. Reed Clark. 2009. Vector-mediated gene transfer engenders long-lived neutralizing activity and protection against SIV infection in monkeys. *Nat Med* 15:901-906.

Jones, S. M., H. Feldmann, U. Stroher, J. B. Geisbert, L. Fernando, A. Grolla, H. D. Klenk, N. J. Sullivan, V. E. Volchkov, E. A. Fritz, K. M. Daddario, L. E. Hensley, P. B. Jahrling, and T. W. Geisbert. 2005. Live attenuated recombinant vaccine protects nonhuman primates against Ebola and Marburg viruses. *Nat Med* 11:786-790.

Kamili, S., V. Sozzi, G. Thompson, K. Campbell, C. M. Walker, S. Locarnini, and K. Krawczynski. 2009. Efficacy of hepatitis B vaccine against antiviral drug-resistant hepatitis B virus mutants in the chimpanzee model. *Hepatology* 49:1483-1491.

Kapikian, A. Z., R. H. Mitchell, R. M. Chanock, R. A. Shvedoff, and C. E. Stewart. 1969. An epidemiologic study of altered clinical reactivity to respiratory syncytial (RS) virus infection in children previously vaccinated with an inactivated RS virus vaccine. *Am J Epidemiol* 89:405-421.

Karron, R. A., P. F. Wright, R. B. Belshe, B. Thumar, R. Casey, F. Newman, F. P. Polack, V. B. Randolph, A. Deatly, J. Hackell, W. Gruber, B. R. Murphy, and P. L. Collins. 2005. Identification of a recombinant live attenuated respiratory syncytial virus vaccine candidate that is highly attenuated in infants. *J Infect Dis* 191:1093-1104.

Keele, B. F., E. E. Giorgi, J. F. Salazar-Gonzalez, J. M. Decker, K. T. Pham, M. G. Salazar, C. Sun, T. Grayson, S. Wang, H. Li, X. Wei, C. Jiang, J. L. Kirchherr, F. Gao, J. A. Anderson, L. H. Ping, R. Swanstrom, G. D. Tomaras, W. A. Blattner, P. A. Goepfert, J. M. Kilby, M. S. Saag, E. L. Delwart, M. P. Busch, M. S. Cohen, D. C. Montefiori, B. F. Haynes, B. Gaschen, G. S. Athreya, H. . Lee, N. Wood, C. Seoighe, A. S. Perelson,

T. Bhattacharya, B. T. Korber, B. H. Hahn, and G. M. Shaw. 2008. Identification and characterization of transmitted and early founder virus envelopes in primary HIV-1 infection. *P Natl Acad Sci USA* 105:7552-7557.

Keele, B. F., J. H. Jones, K. A. Terio, J. D. Estes, R. S. Rudicell, M. L. Wilson, Y. Li, G. H. Learn, T. M. Beasley, J. Schumacher-Stankey, E. Wroblewski, A. Mosser, J. Raphael, S. Kamenya, E. V. Lonsdorf, D. A. Travis, T. Mlengeya, M. J. Kinsel, J. G. Else, G. Silvestri, J. Goodall, P. M. Sharp, G. M. Shaw, A. E. Pusey, and B. H. Hahn. 2009a. Increased mortality and AIDS-like immunopathology in wild chimpanzees infected with SIVcpz. *Nature* 460:515-519.

Keele, B. F., H. Li, G. H. Learn, P. Hraber, E. E. Giorgi, T. Grayson, C. Sun, Y. Chen, W. W. Yeh, N. L. Letvin, J. R. Mascola, G. J. Nabel, B. F. Haynes, T. Bhattacharya, A. S. Perelson, B. T. Korber, B. H. Hahn, and G. M. Shaw. 2009b. Low-dose rectal inoculation of rhesus macaques by SIVsmE660 or SIVmac251 recapitulates human mucosal infection by HIV-1. *J Exp Med* 206:1117-1134.

Khakoo, S. I., C. L. Thio, M. P. Martin, C. R. Brooks, X. Gao, J. Astemborski, J. Cheng, J.J . Goedert, D. Vlahov, M. Hilgartner, S. Cox, A. M. Little, G. J. Alexander, M. E. Cramp, S. J. O'Brien, W. M. Rosenberg, D. L. Thomas, and M. Carrington. 2004. HLA and NK cell inhibitory receptor genes in resolving hepatitis C virus infection. *Science* 305:872-874.

Kim, Y. J., and H. S. Lee. 2010. Increasing incidence of hepatitis A in Korean adults. *Intervirology* 53:10-14.

Kraft, Z., N. R. Derby, R. A. McCaffrey, R. Niec, W. M. Blay, N. L. Haigwood, E. Moysi, C. J. Saunders, T. Wrin, C.J . Petropoulos, M. J. McElrath, and L. Stamatatos. 2007. Macaques infected with a CCR5-tropic simian/human immunodeficiency virus (SHIV) develop broadly reactive anti-HIV neutralizing antibodies. *J Virol* 81:6402-6411.

Krawczynski, K., M. J. Alter, D. L. Tankersley, M. Beach, B. H. Robertson, S. Lambert, G. Kuo, J. E. Spelbring, E. Meeks, S. Sinha, and D. A. Carson. 1996. Effect of immune globulin on the prevention of experimental hepatitis C virus infection. *J Infect Dis* 173:822-828.

Krugman, S. 1986. The Willowbrook hepatitis studies revisited: Ethical aspects. *Rev Infect Dis* 8:157-162.

Krugman, S., and J. P. Giles. 1972. The natural history of viral hepatitis. *Can Med Assoc J* 106 (Suppl):442-446.

Kulkarni, V., R. Jalah, B. Ganneru, C. Bergamaschi, C. Alicea, A. von Gegerfelt, V. Patel, G. M. Zhang, B. Chowdhury, K. E. Broderick, N. Y. Sardesai, A. Valentin, M. Rosati, B. K. Felber, and G. N. Pavlakis. 2010. Comparison of immune responses generated by optimized DNA vaccination against SIV antigens in mice and macaques. *Vaccine* 29:6742-6754.

Kwon, H., and A. S. Lok. 2011. Hepatitis B therapy. *Nat Rev Gastroenterol Hepatol* 8:275-284.

Kwon, S. Y. 2009. Current status of liver diseases in Korea: Hepatitis A. *Korean J Hepatol* 15(Suppl 6):S7-S12.

Laguette, N., B. Sobhian, N. Casartelli, M. Ringeard, C. Chable-Bessia, E. Segeral, A. Yatim, S. Emiliani, O. Schwartz, and M. Benkirane. 2011. SAMHD1 is the dendritic- and myeloid-cell-specific HIV-1 restriction factor counteracted by Vpx. *Nature* 474:654-657.

Landsteiner, K., and C. P. Miller. 1925. Serological studies on the blood of the primates: I. The differentiation of human and anthropoid bloods. *J Exp Med* 42:841-852.

Landsteiner, K., and A. S. Wiener. 1937. On the presence of M agglutinogens in the blood of monkeys. *J Immunol* 33:19-25.

Landsteiner, K., and A. S. Wiener. 1941. Studies on an agglutinogen (Rh) in human blood reacting with anti-rhesus sera and with human isoantibodies. *J Exp Med* 74:309-320.

Lanford, R. E., E. S. Hildebrandt-Eriksen, A. Petri, R. Persson, M. Lindow, M. E. Munk, S. Kauppinen, and H. Orum. 2010. Therapeutic silencing of microRNA-122 in primates with chronic hepatitis C virus infection. *Science* 327:198-201.

Lanford, R. E., Z. Feng, D. Chavez, B. Guerra, K. M. Brasky, Y. Zhou, D. Yamane, A. S. Perelson, C. M. Walker, and S. M. Lemon. 2011. Acute hepatitis A virus infection is associated with a limited type I interferon response and persistence of intrahepatic viral RNA. *P Natl Acad Sci USA* 108:11223-11228.

Lay, M. D., J. Petravic, S. N. Gordon, J. Engram, G. Silvestri, and M. P. Davenport. 2009. Is the gut the major source of virus in early simian immunodeficiency virus infection? *J Virol* 83:7517-7523.

Ledgerwood, J. E., P. Costner, N. Desai, L. Holman, M. E. Enama, G. Yamshchikov, S. Mulangu, Z. Hu, C. A. Andrews, R. A. Sheets, R. A. Koup, M. Roederer, R. Bailer, J. R. Mascola, M. G. Pau, N. J. Sullivan, J. Goudsmit, G. J. Nabel, and B. S. Graham. 2010. A replication defective recombinant Ad5 vaccine expressing Ebola virus GP is safe and immunogenic in healthy adults. *Vaccine* 29:304-313.

Letvin, N. L. 1998. Progress in the development of an HIV-1 vaccine. *Science* 280:1875-1880.

Lewinsohn, D. M., I. S. Tydeman, M. Frieder, J. E. Grotzke, R. A. Lines, S. Ahmed, K. D. Prongay, S. L. Primack, L. M. Colgin, A. D. Lewis, and D. A. Lewinsohn. 2006. High resolution radiographic and fine immunologic definition of TB disease progression in the rhesus macaque. *Microbes Infect* 8:2587-2598.

Li, X., S. Kamili, and K. Krawczynski. 2006. Quantitative detection of hepatitis E virus RNA and dynamics of viral replication in experimental infection. *J Viral Hepat* 13:835-839.

Li, Y., S. A. Migueles, B. Welcher, K. Svehla, A. Phogat, M. K. Louder, X. Wu, G. M. Shaw, M. Connors, R. T. Wyatt, and J. R. Mascola. 2007. Broad

HIV-1 neutralization mediated by CD4-binding site antibodies. *Nat Med* 13:1032-1034.

Lienert, K., and P. Parham. 1996. Evolution of MHC class I genes in higher primates. *Immunol Cell Biol* 74:349-356.

Lifson, J. D., and N. L. Haigwood. In press. Lessons in non-human primate models for AIDS vaccine Research: From minefields to milestones. *In CSH Perspectives*. G. Nabel and F. Bushman, editors. New York: Cold Spring Harbor Press.

Lifson, J. D., M. Piatak, Jr., J. L. Rossio, J. Bess, Jr., E. Chertova, D. Schneider, R. Kiser, V. Coalter, B. Poore, R. Imming, R. C. Desrosiers, L. E. Henderson, and L. O. Arthur. 2002. Whole inactivated SIV virion vaccines with functional envelope glycoproteins: Safety, immunogenicity, and activity against intrarectal challenge. *J Med Primatol* 31:205-216.

Lim, E. S., H. S. Malik, and M. Emerman. 2010. Ancient adaptive evolution of tetherin shaped the functions of Vpu and Nef in human immunodeficiency virus and primate lentiviruses. *J Virol* 84:7124-7134.

Lin, P. L., and J. L. Flynn. 2010. Understanding latent tuberculosis: A moving target. *J Immunol* 185:15-22.

Loffredo, J. T., A. T. Bean, D. R. Beal, E. J. Leon, G. E. May, S. M. Piaskowski, J. R. Furlott, J. Reed, S. K. Musani, E. G. Rakasz, T. C. Friedrich, N. A. Wilson, D. B. Allison, and D. I. Watkins. 2008. Patterns of CD8[+] immunodominance may influence the ability of Mamu-B*08-positive macaques to naturally control simian immunodeficiency virus SIVmac239 replication. *J Virol* 82:1723-1738.

Loffredo, J. T., J. Sidney, A. T. Bean, D. R. Beal, W. Bardet, A. Wahl, O. E. Hawkins, S. Piaskowski, N. A. Wilson, W. H. Hildebrand, D. I. Watkins, and A. Sette. 2009. Two MHC class I molecules associated with elite control of immunodeficiency virus replication, Mamu-B*08 and HLA-B*2705, bind peptides with sequence similarity. *J Immunol* 182:7763-7775.

Mahalanabis, M., P. Jayaraman, T. Miura, F. Pereyra, E. M. Chester, B. Richardson, B Walker, and N. L. Haigwood. 2009. Continuous viral escape and selection by autologous neutralizing antibodies in drug-naive human immunodeficiency virus controllers. *J Virol* 83:662-672.

Malherbe, D. C., N. A. Doria-Rose, L. Misher, T. Beckett, W. B. Puryear, J. T. Schuman, Z. Kraft, J. O'Malley, M. Mori, I. Srivastava, S. Barnett, L. Stamatatos, and N. L. Haigwood. 2011. Sequential immunization with a subtype B HIV-1 envelope quasispecies partially mimics the in vivo development of neutralizing antibodies. *J Virol* 85:5262-5274.

Marthas, M. L., K. K. A. Van Rompay, M. Otsyula, C. J. Miller, D. R. Canfield, N. C. Pedersen, and M. B. McChesney. 1995. Viral factors determine progression to AIDS in simian immunodeficiency virus-infected newborn rhesus macaques. *J Virol* 69:4198-4205.

Mattila, J. T., C. R. Diedrich, P. L. Lin, J. Phuah, and J. L. Flynn. 2011. Simian immunodeficiency virus-induced changes in T cell cytokine responses in cynomolgus macaques with latent Mycobacterium tuberculosis infection are associated with timing of reactivation. *J Immunol* 186:3527-3537.

Maynard, J. E., K. R. Berquist, W. V. Hartwell, and D. H. Krushak. 1972a. Viral hepatitis and studies of hepatitis associated antigen in chimpanzees. *Can Med Assoc J* 106(Suppl):473-479.

Maynard, J. E., K. R. Berquist, D. H. Krushak, and R. H. Purcell. 1972b. Experimental infection of chimpanzees with the virus of hepatitis B. *Nature* 237:514-515.

Maynard, J. E., D. H. Krushak, D. W. Bradley, and K. R. Berquist. 1975. Infectivity studies of hepatitis A and B in non-human primates. *Dev Biol Stand* 30:229-235.

McAleer, W. J., E. B. Buynak, R. Z. Maigetter, D. E. Wampler, W. J. Miller, and M. R. Hilleman. 1984. Human hepatitis B vaccine from recombinant yeast. *Nature* 307:178-180.

McCaustland, K. A., K. Krawczynski, J. W. Ebert, M. S. Balayan, A. G. Andjaparidze, J. E. Spelbring, E. H. Cook, C. Humphrey, P. O. Yarbough, M. O. Favorov, D. Carson, D. W. Bradley, and B. H. Robertson. 2000. Hepatitis E virus infection in chimpanzees: A retrospective analysis. *Arch Virol* 145:1909-1918.

McChesney, M. B., C. J. Miller, P. A. Rota, Y. D. Zhu, L. Antipa, N. W. Lerche, R. Ahmed, and W. J. Bellini. 1997. Experimental measles. I. Pathogenesis in the normal and the immunized host. *Virology* 233:74-84.

McMahan, R. H., L. Golden-Mason, M. I. Nishimura, B. J. McMahon, M. Kemper, T. M. Allen, D. R. Gretch, and H. R. Rosen. 2010. Tim-3 expression on PD-1+ HCV-specific human CTLs is associated with viral persistence, and its blockade restores hepatocyte-directed in vitro cytotoxicity. *J Clin Invest* 120:4546-4557.

Messaoudi, I., J. Warner, M. Fischer, B. Park, B. Hill, J. Mattison, M. A. Lane, G. S. Roth, D. K. Ingram, L. J. Picker, D. C. Douek, M. Mori, and J. Nikolich-Zugich. 2006. Delay of T cell senescence by caloric restriction in aged long-lived nonhuman primates. *P Natl Acad Sci USA* 103:19448-19453.

Messaoudi, I., B. Damania, and S. W. Wong. 2008. Primate models for gammaherpesvirus-associated malignancies. In *DNA Tumor Viruses*. B. Damania and J. Pipas, editors. Springer Science+Business Media LLC.

Messaoudi, I., A. Barron, M. Wellish, F. Engelmann, A. Legasse, S. Planer, D. Gilden, J. Nikolich-Zugich, and R. Mahalingam. 2009. Simian varicella virus infection of rhesus macaques recapitulates essential features of varicella zoster virus infection in humans. *PLoS pathogens* 5:e1000657.

Messaoudi, I., R. Estep, B. Robinson, and S. W. Wong. 2011. Nonhuman primate models of human immunology. *Antioxid Redox Signal* 14:261-273.

Meyer-Olson, D., K. W. Brady, J. T. Blackard, T. M. Allen, S. Islam, N. H. Shoukry, K. Hartman, C. M. Walker, and S. A. Kalams. 2003. Analysis of the TCR beta variable gene repertoire in chimpanzees: Identification of functional homologs to human pseudogenes. *J Immunol* 170:4161-4169.

Meyer-Olson, D., N. H. Shoukry, K. W. Brady, H. Kim, D. P. Olson, K. Hartman, A. K. Shintani, C. M. Walker, and S. A. Kalams. 2004. Limited T cell receptor diversity of HCV-specific T cell responses is associated with CTL escape. *J Exp Med* 200:307-319.

Mikkelsen, T. S., L. W. Hillier, E. E. Eichler, M. C. Zody, D. B. Jaffe, S.-P. Yang, W. Enard, I. Hellmann, K. Lindblad-Toh, T. K. Altheide, N. Archidiacono, P. Bork, J. Butler, J. L. Chang, Z. Cheng, A. T. Chinwalla, P. deJong, K. D. Delehaunty, C. C. Fronick, L. L. Fulton, Y. Gilad, G. Glusman, S. Gnerre, T. A. Graves, T. Hayakawa, K. E. Hayden, X. Huang, H. Ji, W. J. Kent, M.-C. King, E. J. KulbokasIII, M. K. Lee, G. Liu, C. Lopez-Otin, K. D. Makova, O. Man, E. R. Mardis, E. Mauceli, T. L. Miner, W. E. Nash, J. O. Nelson, S. Pääbo, N. J. Patterson, C. S. Pohl, K. S. Pollard, K. Prüfer, X. S. Puente, D. Reich, M. Rocchi, K. Rosenbloom, M. Ruvolo, D. J. Richter, S. F. Schaffner, A. F. A. Smit, S. M. Smith, M. Suyama, J. Taylor, D. Torrents, E. Tuzun, A. Varki, G. Velasco, M. Ventura, J. W. Wallis, M. C. Wendl, R. K. Wilson, E. S. Lander, and R. H. Waterston. 2005. Initial sequence of the chimpanzee genome and comparison with the human genome. *Nature* 437(7055):69-87.

Milush, J. M., D. Kosub, M. Marthas, K. Schmidt, F. Scott, A. Wozniakowski, C. Brown, S. Westmoreland, and D. L. Sodora. 2004. Rapid dissemination of SIV following oral inoculation. *AIDS* 18:2371-2380.

Miyoshi-Akiyama, T., I. Ishida, M. Fukushi, K. Yamaguchi, Y. Matsuoka, T. Ishihara, M. Tsukahara, S. Hatakeyama, N. Itoh, A. Morisawa, Y. Yoshinaka, N. Yamamoto, Z. Lianfeng, Q. Chuan, T. Kirikae, and T. Sasazuki. 2011. Fully human monoclonal antibody directed to proteolytic cleavage site in severe acute respiratory syndrome (SARS) coronavirus S protein neutralizes the virus in a rhesus macaque SARS model. *J Infect Dis* 203:1574-1581.

Mizukoshi, E., M. Nascimbeni, J. B. Blaustein, K. Mihalik, C. M. Rice, T. J. Liang, S. M. Feinstone, and B. Rehermann. 2002. Molecular and immunological significance of chimpanzee major histocompatibility complex haplotypes for hepatitis C virus immune response and vaccination studies. *J Virol* 76:6093-6103.

Monath, T. P., K. R. Brinker, F. W. Chandler, G. E. Kemp, and C. B. Cropp. 1981. Pathophysiologic correlations in a rhesus monkey model of yellow fever with special observations on the acute necrosis of B cell areas of lymphoid tissues. *Am J Trop Med Hyg* 30:431-443.

Moore, P. L., E. S. Gray, D. Sheward, M. Madiga, N. Ranchobe, Z. Lai, W. J. Honnen, M. Nonyane, N. Tumba, T. Hermanus, S. Sibeko, K. Mlisana, S. S. Abdool Karim, C. Williamson, A. Pinter, and L. Morris. 2011. Potent and broad neutralization of HIV-1 subtype C by plasma antibodies targeting a quaternary epitope including residues in the V2 loop. *J Virol* 85:3128-3141.

Moreland, A. J., L. A. Guethlein, R. K. Reeves, K. W. Broman, R. P. Johnson, P. Parham, D. H. O'Connor, and B. N. Bimber. 2011. Characterization of killer immunoglobulin-like receptor genetics and comprehensive genotyping by pyrosequencing in rhesus macaques. *BMC Genomics* 12:295.

Morgan, C., M. Marthas, C. Miller, A. Duerr, C. Cheng-Mayer, R. Desrosiers, J. Flores, N. Haigwood, S. L. Hu, R. P. Johnson, J. Lifson, D. Montefiori, J. Moore, M. Robert-Guroff, H. Robinson, S. Self, and L. Corey. 2008. The use of nonhuman primate models in HIV vaccine development. *PLoS Med* 5:e173.

Mukherjee, S., N. Sarkar-Roy, D. K. Wagener, and P. P. Majumder. 2009. Signatures of natural selection are not uniform across genes of innate immune system, but purifying selection is the dominant signature. *P Natl Acad Sci USA* 106:7073-7078.

Nair, H., D. J. Nokes, B. D. Gessner, M. Dherani, S. A. Madhi, R. J. Singleton, K. L. O'Brien, A. Roca, P. F. Wright, N. Bruce, A. Chandran, E. Theodoratou, A. Sutanto, E. R. Sedyaningsih, M. Ngama, P. K. Munywoki, C. Kartasasmita, E. A. Simoes, I. Rudan, M. W. Weber, and H. Campbell. 2010. Global burden of acute lower respiratory infections due to respiratory syncytial virus in young children: A systematic review and meta-analysis. *Lancet* 375:1545-1555.

Nakamoto, N., H. Cho, A. Shaked, K. Olthoff, M. E. Valiga, M. Kaminski, E. Gostick, D. A. Price, G. J. Freeman, E. J. Wherry, and K. M. Chang. 2009. Synergistic reversal of intrahepatic HCV-specific CD8 T cell exhaustion by combined PD-1/CTLA-4 blockade. *PLoS Pathog* 5:e1000313.

Nara, P. L., W. G. Robey, L. O. Arthur, D. M. Asher, A. V. Wolff, Gibbs, C., Jr., C. D. Gajdusek, and P. J. Fischinger. 1987. Persistent infection of chimpanzees with human immunodeficiency virus: Serological responses and properties of reisolated viruses. *J Virol* 61:3173-3180.

Nguitragool, W., A. A. Bokhari, A. D. Pillai, K. Rayavara, P. Sharma, B. Turpin, L. Aravind, and S. A. Desai. 2011. Malaria parasite clag3 genes determine channel-mediated nutrient uptake by infected red blood cells. *Cell* 145:665-677.

Nichols, H. J. 1914. Observations on experimental typhoid infection of the gall bladder in the rabbit. *J Exp Med* 20:573-581.

Novembre, G. J., M. Saucier, D. C. Anderson, S. A. Klumpp, S. P. O'Neil, C. R. I. Brown, C. E. Hart, P. C. Guenthner, R. B. Swenson, and H. M. McClure. 1997. Development of AIDS in a chimpanzee infected with human immunodeficiency virus type 1. *J Virol* 71:4086-4091.

Okoye, A., H. Park, M. Rohankhedkar, L. Coyne-Johnson, R. Lum, J. M. Walker, S. L. Planer, A. W. Legasse, A. W. Sylwester, M. Piatak, Jr., J. D. Lifson, D. L. Sodora, F. Villinger, M. K. Axthelm, J. E. Schmitz, and L. J. Picker. 2009. Profound $CD4^+/CCR5^+$ T cell expansion is induced by

CD8$^+$ lymphocyte depletion but does not account for accelerated SIV pathogenesis. *J Exp Med* 206:1575-1588.

Olsen, D. B., M. E. Davies, L. Handt, K. Koeplinger, N. R. Zhang, S. W. Ludmerer, D. Graham, N. Liverton, M. MacCoss, D. Hazuda, and S. S. Carroll. 2011. Sustained viral response in a hepatitis C virus-infected chimpanzee via a combination of direct-acting antiviral agents. *Antimicrob Agents Chemother* 55:937-939.

Onlamoon, N. ,S. Noisakran, H. M. Hsiao, A. Duncan, F. Villinger, A. A. Ansari, and G. C. Perng. 2010. Dengue virus-induced hemorrhage in a nonhuman primate model. *Blood* 115:1823-1834.

Otting, N., A. J. de Vos-Rouweler, C. M. Heijmans, N. G. de Groot, G. G. Doxiadis, and R. E. Bontrop. 2007. MHC class I A region diversity and polymorphism in macaque species. *Immunogenetics* 59:367-375.

Pal, R., D. Venzon, N. L. Letvin, S. Santra, D. C. Montefiori, N. R. Miller, E. Tryniszewska, M. G. Lewis, T. C. VanCott, V. Hirsch, R. Woodward, A. Gibson, M. Grace, E. Dobratz, P. D. Markham, Z. Hel, J. Nacsa, M. Klein, J. Tartaglia, and G. Franchini. 2002. ALVAC-SIV-gag-pol-env-based vaccination and macaque major histocompatibility complex class I (A*01) delay simian immunodeficiency virus SIVmac-induced immunodeficiency. *J Virol* 76:292-302.

Pal, R., D. Venzon, S. Santra, V. S. Kalyanaraman, D. C. Montefiori, L. Hocker, L. Hudacik, N. Rose, J. Nacsa, Y. Edghill-Smith, M. Moniuszko, Z. Hel, I. M. Belyakov, J. A. Berzofsky, R. W. Parks, P. D. Markham, N. L. Letvin, J. Tartaglia, and G. Franchini. 2006. Systemic immunization with an ALVAC-HIV-1/protein boost vaccine strategy protects rhesus macaques from CD4$^+$ T-cell loss and reduces both systemic and mucosal simian-human immunodeficiency virus SHIVKU2 RNA levels. *J Virol* 80:3732-3742.

Pandrea, I., C. Apetrei, J. Dufour, N. Dillon, J. Barbercheck, M. Metzger, B. Jacquelin, R. Bohm, P. A. Marx, F. Barre-Sinoussi, V. M. Hirsch, M. C. Muller-Trutwin, A. A. Lackner, and R. S. Veazey. 2006. Simian immunodeficiency virus SIVagm.sab infection of Caribbean African green monkeys: a new model for the study of SIV pathogenesis in natural hosts. *J Virol* 80:4858-4867.

Pandrea, I., C. Apetrei, S. Gordon, J. Barbercheck, J. Dufour, R. Bohm, B. Sumpter, P. Roques, P. A. Marx, V. M. Hirsch, A. Kaur, A. A. Lackner, R. S. Veazey, and G. Silvestri. 2007. Paucity of CD4$^+$CCR5$^+$ T cells is a typical feature of natural SIV hosts. *Blood* 109:1069-1076.

Parham, P. 2005. MHC class I molecules and KIRs in human history, health and survival. *Nature Rev Immunol* 5:201-214.

Parham, P. 2008. The genetic and evolutionary balances in human NK cell receptor diversity. *Semin Immunol* 20:311-316.

Parham, P., L. Abi-Rached, L. Matevosyan, A. K. Moesta, P. J. Norman, A. M. Older Aguilar, and L. A. Guethlein. 2010. Primate-specific regulation of natural killer cells. *Journal Med Primatol* 39:194-212.

Patel, V., A. Valentin, V. Kulkarni, M. Rosati, C. Bergamaschi, R. Jalah, C. Alicea, J. T. Minang, M. T. Trivett, C. Ohlen, J. Zhao, M. Robert-Guroff, A. S. Khan, R. Draghia-Akli, B. K. Felber, and G. N. Pavlakis. 2010. Long-lasting humoral and cellular immune responses and mucosal dissemination after intramuscular DNA immunization. *Vaccine* 28:4827-4836.

Patterson, L. J., N. Malkevitch, J. Pinczewski, D. Venzon, Y. Lou, B. Peng, C. Munch, M. Leonard, E. Richardson, K. Aldrich, V. S. Kalyanaraman, G. N. Pavlakis, and M. Robert-Guroff. 2003. Potent, persistent induction and modulation of cellular immune responses in rhesus macaques primed with Ad5hr-simian immunodeficiency virus (SIV) env/rev, gag, and/or nef vaccines and boosted with SIV gp120. *J Virol* 77:8607-8620.

Penna, A., M. Pilli, A. Zerbini, A. Orlandini, S. Mezzadri, L. Sacchelli, G. Missale, and C. Ferrari. 2007. Dysfunction and functional restoration of HCV-specific CD8 responses in chronic hepatitis C virus infection. *Hepatology* 45:588-601.

Permar, S. R., S. A. Klumpp, K. G. Mansfield, W. K. Kim, D. A. Gorgone, M. A. Lifton, K. C. Williams, J. E. Schmitz, K. A. Reimann, M. K. Axthelm, F. P. Polack, D. E. Griffin, and N. L. Letvin. 2003. Role of CD8(+) lymphocytes in control and clearance of measles virus infection of rhesus monkeys. *J Virol* 77:4396-4400.

Permar, S. R., S. A. Klumpp, K. G. Mansfield, A. A. Carville, D. A. Gorgone, M. A. Lifton, J. E. Schmitz, K. A. Reimann, F. P. Polack, D. E. Griffin, and N. L. Letvin. 2004. Limited contribution of humoral immunity to the clearance of measles viremia in rhesus monkeys. *J Infect Dis* 190:998-1005.

Permar, S. R., S. S. Rao, Y. Sun, S. Bao, A. P. Buzby, H. H. Kang, and N. L. Letvin. 2007. Clinical measles after measles virus challenge in simian immunodeficiency virus-infected measles virus-vaccinated rhesus monkeys. *J Infect Dis* 196:1784-1793.

Pitcher, C. J., S. I. Hagen, J. M. Walker, R. Lum, B. L. Mitchell, V. C. Maino, M. K. Axthelm, and L. J. Picker. 2002. Development and homeostasis of T cell memory in rhesus macaque. *J Immunol* 168:29-43.

Polacino, P., K. Larsen, L. Galmin, J. Suschak, Z. Kraft, L. Stamatatos, D. Anderson, S. W. Barnett, R. Pal, K. Bost, A. H. Bandivdekar, C. J. Miller, and S. L. Hu. 2008. Differential pathogenicity of SHIV infection in pig-tailed and rhesus macaques. *J Med Primatol* 37(Suppl 2):13-23.

Poland, J. D., C. H. Calisher, T. P. Monath, W. G. Downs, and K. Murphy. 1981. Persistence of neutralizing antibody 30-35 years after immunization with 17D yellow fever vaccine. *B World Health Org* 59:895-900.

Porter, D. W., F. M. Thompson, T. K. Berthoud, C. L. Hutchings, L. Andrews, S. Biswas, I. Poulton, E. Prieur, S. Correa, R. Rowland, T. Lang, J. Williams, S. C. Gilbert, R. E. Sinden, S. Todryk, and A. V. Hill. 2011. A human Phase I/IIa malaria challenge trial of a polyprotein malaria vaccine. *Vaccine* 29:7514-7522.

Price, D. A., A. D. Bitmansour, J. B. Edgar, J. M. Walker, M. K. Axthelm, D. C. Douek, and L. J. Picker. 2008. Induction and evolution of cytomegalovirus-specific CD4$^+$ T cell clonotypes in rhesus macaques. *J Immunol* 180:269-280.

Prince, A. M., and L. Andrus. 1998. AIDS vaccine trials in chimpanzees. *Science* 282:2195-2196.

Prince, A. M., B. Brotman, G. F. Grady, W. J. Kuhns, C. Hazzi, R. W. Levine, and S. J. Millian. 1974. Long-incubation post-transfusion hepatitis without serological evidence of exposure to hepatitis-B virus. *Lancet* 2:241-246.

Prince, A. M., D. Pascual, L. B. Kosolapov, D. Kurokawa, L. Baker, and P. Rubinstein. 1987. Prevalence, clinical significance, and strain specificity of neutralizing antibody to the human immunodeficiency virus. *J Infect Dis* 156:268-272.

Prince, A. M., H. Reesink, D. Pascual, B. Horowitz, I. Hewlett, K. K. Murthy, K. E. Cobb, and J. Eichberg. 1991. Prevention of HIV infection by passive immunization with HIV immunoglobulin. *AIDS Res Hum Retrov* 7:971-973.

Prince, A. M., B. Brotman, T. Huima, D. Pascual, M. Jaffery, and G. Inchauspe. 1992. Immunity in hepatitis C infection. *J Infect Dis* 165:438-443.

Provost, P. J., P. A. Conti, P. A. Giesa, F. S. Banker, E. B. Buynak, W. J. McAleer, and M. R. Hilleman. 1983. Studies in chimpanzees of live, attenuated hepatitis A vaccine candidates. *Proc Soc Exp Biol Med* 172:357-363.

Pulendran, B., J. Miller, T. D. Querec, R. Akondy, N. Moseley, O. Laur, J. Glidewell, N. Monson, T. Zhu, H. Zhu, S. Staprans, D. Lee, M. A. Brinton, A. A. Perelygin, C. Vellozzi, P. Brachman, Jr., S. Lalor, D. Teuwen, R. B. Eidex, M. Cetron, F. Priddy, C. del Rio, J. Altman, and R. Ahmed. 2008. Case of yellow fever vaccine–associated viscerotropic disease with prolonged viremia, robust adaptive immune responses, and polymorphisms in CCR5 and RANTES genes. *J Infect Dis* 198:500-507.

Purcell, R. H., and J. L. Gerin. 1975. Hepatitis B subunit vaccine: A preliminary report of safety and efficacy tests in chimpanzees. *Am J Med Sci* 270:395-399.

Purcell, R. H., E. D'Hondt, R. Bradbury, S. U. Emerson, S. Govindarajan, and L. Binn. 1992. Inactivated hepatitis A vaccine: Active and passive immunoprophylaxis in chimpanzees. *Vaccine* 10(Suppl 1):S148-S151.

Querec, T. D., R. S. Akondy, E. K. Lee, W. Cao, H. I. Nakaya, D. Teuwen, A. Pirani, K. Gernert, J. Deng, B. Marzolf, K. Kennedy, H. Wu, S. Bennouna, H. Oluoch, J. Miller, R.Z. Vencio, M. Mulligan, A. Aderem, R. Ahmed, and B. Pulendran. 2009. Systems biology approach predicts immunogenicity of the yellow fever vaccine in humans. *Nat Immunol* 10:116-125.

Raziorrouh, B., W. Schraut, T. Gerlach, D. Nowack, N. H. Gruner, A. Ulsenheimer, R. Zachoval, M. Wachtler, M. Spannagl, J. Haas, H. M. Diepolder, and M. C. Jung. 2010. The immunoregulatory role of CD244 in chronic hepatitis B infection and its inhibitory potential on virus-specific CD8[+] T-cell function. *Hepatology* 52:1934-1947.

Reeves, R. K., J. Gillis, F. E. Wong, Y. Yu, M. Connole, and R. P. Johnson. 2010. CD16- natural killer cells: Enrichment in mucosal and secondary lymphoid tissues and altered function during chronic SIV infection. *Blood* 115:4439-4446.

Reeves, R. K., P. A. Rajakumar, T. I. Evans, M. Connole, J. Gillis, F. E. Wong, Y. V. Kuzmichev, A. Carville, and R. P. Johnson. 2011. Gut inflammation and indoleamine deoxygenase inhibit IL-17 production and promote cytotoxic potential in NKp44+ mucosal NK cells during SIV infection. *Blood* 118:3321-3330.

Rehermann, B. 2009. Hepatitis C virus versus innate and adaptive immune responses: A tale of coevolution and coexistence. *J Clin Invest* 119:1745-1754.

Reimann, K. A., J. T. Li, R. Veazey, M. Halloran, I. W. Park, G. B. Karlsson, J. Sodroski, and N. L. Letvin. 1996. A chimeric simian/human immunodeficiency virus expressing a primary patient human immunodeficiency virus type 1 isolate env causes an AIDS-like disease after in vivo passage in rhesus monkeys. *J Virol* 70:6922-6928.

Rerks-Ngarm, S., P. Pitisuttithum, S. Nitayaphan, J. Kaewkungwal, J. Chiu, R. Paris, N. Premsri, C. Namwat, M. de Souza, E. Adams, M. Benenson, S. Gurunathan, J. Tartaglia, J. G. McNeil, D. P. Francis, D. Stablein, D. L. Birx, S. Chunsuttiwat, C. Khamboonruang, P. Thongcharoen, M. L. Robb, N. L. Michael, P. Kunasol, and J. H. Kim. 2009. Vaccination with ALVAC and AIDSVAX to prevent HIV-1 infection in Thailand. *N Engl J Med* 361:2209-2220.

Rhee, E. G., and D. H. Barouch. 2009. Translational Mini-Review Series on Vaccines for HIV: Harnessing innate immunity for HIV vaccine development. *Clin Exp Immunol* 157:174-180.

Rijckborst, V., M. J. Sonneveld, and H. L. Janssen. 2011. Review article: Chronic hepatitis B—anti-viral or immunomodulatory therapy? *Aliment Pharmacol Ther* 33:501-513.

Rizzetto, M., M. G. Canese, R. H. Purcell, W. T. London, L. D. Sly, and J. L. Gerin. 1981. Experimental HBV and delta infections of chimpanzees: Occurrence and significance of intrahepatic immune complexes of HBcAg and delta antigen. *Hepatology* 1:567-574.

Robinson, J. E., K. S. Cole, D. H. Elliott, H. Lam, A. M. Amedee, R. Means, R. C. Desrosiers, J. Clements, R. C. Montelaro, and M. Murphey-Corb. 1998. Production and characterization of SIV envelope-specific rhesus monoclonal antibodies from a macaque asymptomatically infected with a live SIV vaccine. *AIDS Res Hum Retrov* 14:1253-1262.

Rosati, M., A. von Gegerfelt, P. Roth, C. Alicea, A. Valentin, M. Robert-Guroff, D. Venzon, D. C. Montefiori, P. Markham, B. K. Felber, and G. N. Pavlakis. 2005. DNA vaccines expressing different forms of simian immunodeficiency virus antigens decrease viremia upon SIVmac251 challenge. *J Virol* 79:8480-8492.

Rudensey, L. M., J. T. Kimata, E. M. Long, B. Chackerian, and J. Overbaugh. 1998. Changes in the extracellular envelope glycoprotein of variants that evolve during the course of simian immunodeficiency virus SIVMne infection affect neutralizing antibody recognition, syncytium formation and macrophage tropism, but not replication, cytopathicity or CCR-5 coreceptor recognition. *J Virol* 72:209-217.

Sa, J. M., O. Twu, K. Hayton, S. Reyes, M. P. Fay, P. Ringwald, and T. E. Wellems. 2009. Geographic patterns of Plasmodium falciparum drug resistance distinguished by differential responses to amodiaquine and chloroquine. *P Natl Acad Sci USA* 106:18883-18889.

Sabatini, M. J., P. Ebert, D. A. Lewis, P. Levitt, J. L. Cameron, and K. Mirnics. 2007. Amygdala gene expression correlates of social behavior in monkeys experiencing maternal separation. *J Neurosci* 27:3295-3304.

Sambrook, J. G., A. Bashirova, H. Andersen, M. Piatak, G. S. Vernikos, P. Coggill, J. D. Lifson, M. Carrington, and S. Beck. 2006. Identification of the ancestral killer immunoglobulin-like receptor gene in primates. *BMC Genomics* 7:209.

Sarma, S. V., U. T. Eden, M. L. Cheng, Z. M. Williams, R. Hu, E. Eskandar, and E. N. Brown. 2010. Using point process models to compare neural spiking activity in the subthalamic nucleus of Parkinson's patients and a healthy primate. *IEEE T Bio-med Eng* 57:1297-1305.

Sather, D. N., J. Armann, L. K. Ching, A. Mavrantoni, G. Sellhorn, Z. Caldwell, X. Yu, B. Wood, S. Self, S. Kalams, and L. Stamatatos. 2009. Factors associated with the development of cross-reactive neutralizing antibodies during human immunodeficiency virus type 1 infection. *J Virol* 83:757-769.

Sauter, D., M Schindler, A. Specht, W. N. Landford, J. Munch, K. A. Kim, J. Votteler, U. Schubert, F. Bibollet-Ruche, B. F. Keele, J. Takehisa, Y. Ogando, C. Ochsenbauer, J. C. Kappes, A. Ayouba, M. Peeters, G. H. Learn, G. Shaw, P. M. Sharp, P. Bieniasz, B. H. Hahn, T. Hatziioannou, and F. Kirchhoff. 2009. Tetherin-driven adaptation of Vpu and Nef function and the evolution of pandemic and nonpandemic HIV-1 strains. *Cell Host Microbe* 6:409-421.

Schatten, G., and S. Mitalipov. 2009. Developmental biology: Transgenic primate offspring. *Nature* 459:515-516.

Schmitz, J. E., M. J. Kuroda, S. Santra, V. G. Sasseville, M. A. Simon, M. A. Lifton, P. Racz, K. Tenner-Racz, M. Dalesandro, B. J. Scallon, J. Ghrayeb, M. A. Forman, D. C. Montefiori, E. P. Rieber, N. L. Letvin, and K. A. Reimann. 1999. Control of viremia in simian immunodeficiency virus infection by $CD8^+$ lymphocytes. *Science* 283:857-860.

Schmitz, J. E., M. J. Kuroda, S. Santra, M. A. Simon, M. A. Lifton, W. Lin, R. Khunkhun, M. Piatak, J. D. Lifson, G. Grosschupff, R. S. Gelman, P. Racz, K. Tenner-Racz, K. A. Mansfield, N. L. Letvin, D. C. Montefiori, and K. A. Reimann. 2003. Effect of humoral immune responses on controlling viremia during primary infection of rhesus monkeys with simian immunodeficiency virus. *J Virol* 77:2165-2173.

Schmokel, J., H. Li, E. Bailes, M. Schindler, G. Silvestri, B. H. Hahn, C. Apetrei, and F. Kirchhoff. 2009. Conservation of Nef function across highly diverse lineages of SIVsmm. *Retrovirology* 6:36.

Scinicariello, F., C. N. Engleman, L. Jayashankar, H. M. McClure, and R. Attanasio. 2004. Rhesus macaque antibody molecules: Sequences and heterogeneity of alpha and gamma constant regions. *Immunology* 111:66-74.

Shiver, J. W., T. M. Fu, L. Chen, D. R. Casimiro, M. E. Davies, R. K. Evans, Z. Q. Zhang, A. J. Simon, W. L. Trigona, S. A. Dubey, L. Huang, V. A. Harris, R. S. Long, X. Liang, L. Handt, W. A. Schleif, L. Zhu, D. C. Freed, N. V. Persaud, L. Guan, K. S. Punt, A. Tang, M. Chen, K. A. Wilson, K. B. Collins, G. J. Heidecker, V. R. Fernandez, H. C. Perry, J. G. Joyce, K. M. Grimm, J. C. Cook, P. M. Keller, D. S. Kresock, H. Mach, R. D. Troutman, L. A. Isopi, D. M. Williams, Z. Xu, K. E. Bohannon, D. B. Volkin, D. C. Montefiori, A. Miura, G. R. Krivulka, M. A. Lifton, M. J. Kuroda, J. E. Schmitz, N. L. Letvin, M. J. Caulfield, A. J. Bett, R. Youil, D. C. Kaslow, and E. A. Emini. 2002. Replication-incompetent adenoviral vaccine vector elicits effective anti-immunodeficiency-virus immunity. *Nature* 415:331-335.

Shoukry, N. H., A. Grakoui, M. Houghton, D. Y. Chien, J. Ghrayeb, K. A. Reimann, and C. M. Walker. 2003. Memory CD8[+] T cells are required for protection from persistent hepatitis C virus infection. *J Exp Med* 197:1645-1655.

Shrestha, M. P., R. M. Scott, D. M. Joshi, M. P. Mammen, Jr., G. B. Thapa, N. Thapa, K. S. Myint, M. Fourneau, R. A. Kuschner, S. K. Shrestha, M. P. David, J. Seriwatana, D. W. Vaughn, A. Safary, T. P. Endy, and B. L. Innis. 2007. Safety and efficacy of a recombinant hepatitis E vaccine. *N Engl J Med* 356:895-903.

Sidney, J., S. Asabe, B. Peters, K. A. Purton, J. Chung, T. J. Pencille, R. Purcell, C. M. Walker, F. V. Chisari, and A. Sette. 2006. Detailed characterization of the peptide binding specificity of five common Patr class I MHC molecules. *Immunogenetics* 58:559-570.

Silvestri, G. 2009. Immunity in natural SIV infections. *J Intern Med* 265:97-109.

Smirnova, I., A. Poltorak, E. K. Chan, C. McBride, and B. Beutler. 2000. Phylogenetic variation and polymorphism at the toll-like receptor 4 locus (TLR4). *Genome Biol* 1:RESEARCH002.

Smit-McBride, Z., J. J. Mattapallil, M. McChesney, D. Ferrick, and S. Dandekar. 1998. Gastrointestinal T lymphocytes retain high potential for cytokine responses but have severe CD4(+) T-cell depletion at all stages of simian immunodeficiency virus infection compared to peripheral lymphocytes. *J Virol* 72:6646-6656.

Smith, L. R., M. K. Wloch, M. Ye, L. R. Reyes, S. Boutsaboualoy, C. E. Dunne, J. A. Chaplin, D. Rusalov, A. P. Rolland, C. L. Fisher, M. S. Al-Ibrahim, M. L. Kabongo, R. Steigbigel, R. B. Belshe, E. R. Kitt, A. H. Chu, and R. B. Moss. 2010. Phase 1 clinical trials of the safety and immunogenicity of adjuvanted plasmid DNA vaccines encoding influenza A virus H5 hemagglutinin. *Vaccine* 28:2565-2572.

Smits, S. L., A. de Lang, J. M. van den Brand, L. M. Leijten, I. W. F. van, M. J. Eijkemans, G. van Amerongen, T. Kuiken, A. C. Andeweg, A. D. Osterhaus, and B. L. Haagmans. 2010. Exacerbated innate host response to SARS-CoV in aged non-human primates. *PLoS pathogens* 6:e1000756.

Soderstrom, K., J. O'Malley, K. Steece-Collier, and J. H. Kordower. 2006. Neural repair strategies for Parkinson's disease: Insights from primate models. *Cell Transplant* 15:251-265.

Sodora, D. L., C. J. Miller, and P. A. Marx. 1997. Vaginal transmission of SIV: assessing infectivity and hormonal influences in macaques inoculated with cell-free and cell-associated viral stocks. *AIDS Res Hum Retrov* 13:S1-S5.

Sodora, D. L., J. S. Allan, C. Apetrei, J. M. Brenchley, D. C. Douek, J. G. Else, J. D. Estes, B. H. Hahn, V. M. Hirsch, A. Kaur, F. Kirchhoff, M. Muller-Trutwin, I. Pandrea, J. E. Schmitz, and G. Silvestri. 2009. Toward an AIDS vaccine: Lessons from natural simian immunodeficiency virus infections of African nonhuman primate hosts. *Nat Med* 15:861-865.

St Gelais, C., and L. Wu. 2011. SAMHD1: A new insight into HIV-1 restriction in myeloid cells. *Retrovirology* 8:55.

Stevens, H. E., J. F. Leckman, J. D. Coplan, and S. J. Suomi. 2009. Risk and resilience: Early manipulation of macaque social experience and persistent behavioral and neurophysiological outcomes. *J Am Acad Child Ps* 48:114-127.

Sullivan, N. J., A. Sanchez, P. E. Rollin, Z. Y. Yang, and G. J. Nabel. 2000. Development of a preventive vaccine for Ebola virus infection in primates. *Nature* 408:605-609.

Sullivan, N. J., T. W. Geisbert, J. B. Geisbert, L. Xu, Z. Y. Yang, M. Roederer, R. A. Koup, P. B. Jahrling, and G. J. Nabel. 2003. Accelerated vaccination for Ebola virus haemorrhagic fever in non-human primates. *Nature* 424:681-684.

Sullivan, N. J., T. W. Geisbert, J. B. Geisbert, D. J. Shedlock, L. Xu, L. Lamoreaux, J. H. Custers, P. M. Popernack, Z. Y. Yang, M. G. Pau, M. Roederer, R. A. Koup, J. Goudsmit, P. B. Jahrling, and G. J. Nabel. 2006. Immune protection of nonhuman primates against Ebola virus with single low-dose adenovirus vectors encoding modified GPs. *PLoS Med* 3:e177.

Sullivan, N. J., J. E. Martin, B. S. Graham, and G. J. Nabel. 2009. Correlates of protective immunity for Ebola vaccines: Implications for regulatory approval by the animal rule. *Nature Rev Micro* 7:393-400.

Tabor, E., R. J. Gerety, J. A. Drucker, L. B. Seeff, J.H. Hoofnagle, D. R. Jackson, M. April, L. F. Barker, and G. Pineda-Tamondong. 1978. Transmission of non-A, non-B hepatitis from man to chimpanzee. *Lancet* 1:463-466.

Tachibana, M., M. Sparman, H. Sritanaudomchai, H. Ma, L. Clepper, J. Woodward, Y. Li, C. Ramsey, O. Kolotushkina, and S. Mitalipov. 2009. Mitochondrial gene replacement in primate offspring and embryonic stem cells. *Nature* 461:367-372.

Teng, M. N., S. S. Whitehead, A. Bermingham, M. St Claire, W. R. Elkins, B. R. Murphy, and P. L. Collins. 2000. Recombinant respiratory syncytial virus that does not express the NS1 or M2-2 protein is highly attenuated and immunogenic in chimpanzees. *J Virol* 74:9317-9321.

Thimme, R., D. Oldach, K. M. Chang, C. Steiger, S. C. Ray, and F. V. Chisari. 2001. Determinants of viral clearance and persistence during acute hepatitis C virus infection. *J Exp Med* 194:1395-1406.

Thimme, R., J. Bukh, H. C. Spangenberg, S. Wieland, J. Pemberton, C. Steiger, S. Govindarajan, R. H. Purcell, and F. V. Chisari. 2002. Viral and immunological determinants of hepatitis C virus clearance, persistence, and disease. *P Natl Acad Sci USA* 99:15661-15668.

Thimme, R., S. Wieland, C. Steiger, J. Ghrayeb, K. A. Reimann, R. H. Purcell, and F. V. Chisari. 2003. CD8(+) T cells mediate viral clearance and disease pathogenesis during acute hepatitis B virus infection. *J Virol* 77:68-76.

Ulmer, J. B., J. J. Donnelly, S. E. Parker, G. H. Rhodes, P. L. Felgner, V. J. Dwarki, S. H. Gromkowski, R. R. Deck, C. M. EdWitt, A. Friedman, L. A. Hawe, K. R. Leander, D. Mrtinex, H. C. Perry, J. W. Shiver, D. L. Mongomery, and M. A. Liu. 1993. Heterologous protection against influenza by injection of DNA encoding a viral protein. *Science* 259:1745-1749.

Vaine, M., S. Wang, E. T. Crooks, P. Jiang, D. C. Montefiori, J. Binley, and S. Lu. 2008. Improved induction of antibodies against key neutralizing epitopes by human immunodeficiency virus type 1 gp120 DNA prime-protein boost vaccination compared to gp120 protein-only vaccination. *J Virol* 82:7369-7378.

van Montfort, T., M. Melchers, G. Isik, S. Menis, P. S. Huang, K. Matthews, E. Michael, B. Berkhout, W. R. Schief, J. P. Moore, and R. W. Sanders. 2011. A chimeric HIV-1 envelope glycoprotein trimer with an embedded granulocyte-macrophage colony-stimulating factor (GM-CSF) domain induces enhanced antibody and T cell responses. *J Biol Chem* 286:22250-22261.

Van Rompay, K. K. A. , and N. L. Haigwood. 2008. Pediatric AIDS: Maternal-fetal and maternal-infant transmission of lentiviruses and effects on infant development in nonhuman primates. In *Primate Models of Children's Health and Developmental Disabilities.* T. M. Burbacher, G. P. Sackett, and K. S. Grant, editors. Amsterdam: Academic Press.

VandeBerg, J. L., S. M. Zola, J. J. Ely, and R. C. Kennedy. 2006. Monoclonal antibody testing. *J Med Primatol* 35:405-406.

Vierboom, M., and B. 't Hart. 2008. Spontaneous and experimentally induced autoimmune diseases in nonhuman primates. In *Primate Models of Children's Health and Developmental Disabilities.* T. Burbacher, G. Sackett, and K. Grant, editors. New York: Academic Press. Pp. 71-107.

Voytko, M. L., and G. P. Tinkler. 2004. Cognitive function and its neural mechanisms in nonhuman primate models of aging, Alzheimer disease, and menopause. *Frontiers in Bioscience: A Journal and Virtual Library* 9:1899-1914.

Walker, C. M. 2010. Adaptive immunity to the hepatitis C virus. *Adv Virus Res* 78:43-86.

Walker, J. M., H. T. Maecker, V. C. Maino, and L. J. Picker. 2004. Multicolor flow cytometric analysis in SIV-infected rhesus macaque. *Meth Cell Bio* 75:535-557.

Walker, L. M., M. D. Simek, F. Priddy, J. S. Gach, D. Wagner, M. B. Zwick, S. K. Phogat, P. Poignard, and D. R. Burton. 2010. A limited number of antibody specificities mediate broad and potent serum neutralization in selected HIV-1 infected individuals. *PLoS pathogens* 6:e1001028.

Warfield, K. L., D. L. Swenson, G. G. Olinger, W. V. Kalina, M. J. Aman, and S. Bavari. 2007a. Ebola virus-like particle-based vaccine protects nonhuman primates against lethal Ebola virus challenge. *J Infect Dis* 196 (Suppl 2):S430-S437.

Warfield, K. L., D. L. Swenson, G. G. Olinger, W. V. Kalina, M. Viard, M. Aitichou, X. Chi, S. Ibrahim, R. Blumenthal, Y. Raviv, S. Bavari, and M. J. Aman. 2007b. Ebola virus inactivation with preservation of antigenic and structural integrity by a photoinducible alkylating agent. *J Infect Dis* 196(Suppl 2):S276-S283.

Wedemeyer, H., and S. Pischke. 2011. Hepatitis: Hepatitis E vaccination—is HEV 239 the breakthrough? *Nat Rev Gastroenterol Hepatol* 8:8-10.

Whitehead, S. S., M. G. Hill, C. Y. Firestone, M. St Claire, W. R. Elkins B. R. Murphy, and P. L. Collins. 1999. Replacement of the F and G proteins of respiratory syncytial virus (RSV) subgroup A with those of subgroup B generates chimeric live attenuated RSV subgroup B vaccine candidates. *J Virol* 73:9773-9780.

Willey, R. L., R. Byrum, M. Piatak, Y. B. Kim, M. W. Cho, J. L. Rossio Jr., J. Bess Jr., T. Igarashi, Y. Endo, L. O. Arthur, J. D. Lifson, and M. A. Martin. 2003. Control of viremia and prevention of simian-human immunodeficiency virus-induced disease in rhesus macaques immunized with recombinant vaccinia viruses plus inactivated simian immunodeficiency virus and human immunodeficiency virus type 1 particles. *J Virol* 77:1163-1174.

Wilson, S. J., B. L. Webb, C. Maplanka, R. M. Newman, E. J. Verschoor, J. L. Heeney, and G. J. Towers. 2008. Rhesus macaque TRIM5 alleles have divergent antiretroviral specificities. *J Virol* 82:7243-7247.

Wooding, S., A. C. Stone, D. M. Dunn, S. Mummidi, L. B. Jorde, R. K. Weiss, S. Ahuja, and M. J. Bamshad. 2005. Contrasting effects of natural selection on human and chimpanzee CC chemokine receptor 5. *Am J Hum Genet* 76:291-301.

Wu, X., Z. Y. Yang, Y. Li, C. M. Hogerkorp, W. R. Schief, M. S. Seaman, T. Zhou, S. D. Schmidt, L. Wu, L. Xu, N. S. Longo, K. McKee, S. O'Dell, M. K. Louder, D. L. Wycuff, Y. Feng, M. Nason, N. Doria-Rose, M. Connors, P. D. Kwong, M. Roederer, R. T. Wyatt, G. J. Nabel, and J. R. Mascola. 2010. Rational design of envelope identifies broadly neutralizing human monoclonal antibodies to HIV-1. *Science* 329:856-861.

Yant, L. J., T. C. Friedrich, R. C. Johnson, G. E. May, N. J. Maness, A. M. Enz, J. D. Lifson, D. H. O'Connor, M. Carrington, and D. I. Watkins. 2006. The high-frequency major histocompatibility complex class I allele Mamu-B*17 is associated with control of simian immunodeficiency virus SIVmac239 replication. *J Virol* 80:5074-5077.

Yeh, W. W., P. Jaru-Ampornpan, D. Nevidomskyte, M. Asmal, S. S. Rao, A. P. Buzby, D. C. Montefiori, B. T. Korber, and N. L. Letvin. 2009. Partial protection of Simian immunodeficiency virus (SIV)-infected rhesus monkeys against superinfection with a heterologous SIV isolate. *J Virol* 83:2686-2696.

Yin, J., A. Dai, J. Lecureux, T. Arango, M. A. Kutzler, J. Yan, M. G. Lewis, A. Khan, N. Y. Sardesai, D. Montefiore, R. Ruprecht, D. B. Weiner, and J. D. Boyer. 2010. High antibody and cellular responses induced to HIV-1 clade C envelope following DNA vaccines delivered by electroporation. *Vaccine* 29:6763-6770.

Yu, C., D. Boon, S. L. McDonald, T. G. Myers, K. Tomioka, H. Nguyen, R. E. Engle, S. Govindarajan, S. U. Emerson, and R. H. Purcell. 2010a. Pathogenesis of hepatitis E virus and hepatitis C virus in chimpanzees: Similarities and differences. *J Virol* 84:11264-11278.

Yu, C., D. Boon, S. L. McDonald, T. G. Myers, K. Tomioka, H. Nguyen, R. E. Engle, S. Govindarajan, S. U. Emerson, and R. H. Purcell. 2010b. Pathogenesis of hepatitis E virus and hepatitis C virus in chimpanzees: Similarities and differences. *J Virol* 84:11264-11278.

Zhu, Y. D., J. Heath, J. Collins, T. Greene, L. Antipa, P. Rota, W. Bellini, and M. McChesney. 1997. Experimental measles. II. Infection and immunity in the rhesus macaque. *Virology* 233:85-92.

Zhu, F. C., J. Zhang, X. F. Zhang, C. Zhou, Z. Z. Wang, S. J. Huang, H. Wang, C. L. Yang, H. M. Jiang, J. P. Cai, Y. J. Wang, X. Ai, Y. M. Hu, Q. Tang, X. Yao, Q. Yan, Y. L. Xian, T. Wu, Y. M. Li, J. Miao, M. H. Ng, J. W. Shih, and N. S. Xia. 2010. Efficacy and safety of a recombinant hepatitis E vaccine in healthy adults: A large-scale, randomised, double-blind placebo-controlled, phase 3 trial. *Lancet* 376:895-902.

Zolla-Pazner, S., X. P. Kong, X. Jiang, T. Cardozo, A. Nadas, S. Cohen, M. Totrov, M. S. Seaman, S. Wang, and S. Lu. 2011. Cross-clade HIV-1 neutralizing antibodies induced with V3-scaffold protein immunogens following priming with gp120 DNA. *J Virol.*

C

Information-Gathering Agendas

May 26, 2011

Keck Center, Room 109
500 Fifth Street, NW
Washington, DC 20001

BACKGROUND AND OVERVIEW

Session Objectives: Obtain a better understanding of the background to the study and the charge to the committee. Receive a briefing from NIH about existing areas of science where chimpanzee research is supported. Hear from stakeholders about the use of chimpanzees in research, as specifically related to the committee's charge.

1:00 p.m.   Welcome and Introductions

JOHN STOBO, *Committee Chair*
Senior Vice President
Health Sciences and Services
University of California System

1:10 p.m.   Background and Charge to the Committee

SALLY ROCKEY
Deputy Director for Extramural Research
National Institutes of Health

1:30 p.m.   Committee Discussion with Sponsor

                   JOHN STOBO, *Committee Chair*
                   Senior Vice President
                   Health Sciences and Services
                   University of California System

2:15 p.m.   NIH-Supported Chimpanzee Biomedical Research

                   HAROLD WATSON
                   Deputy Director
                   Division of Comparative Medicine
                   National Center for Research Resources, NIH

2:35 p.m.   Discussion with the Committee

2:45 p.m.   BREAK

3:15 p.m.   NIH-Supported Chimpanzee Behavioral Research

                   RICHARD NAKAMURA
                   Scientific Director
                   National Institute of Mental Health, NIH

3:35 p.m.   Discussion with the Committee

3:45 p.m.   Panel Discussion: Is there a continued need for
                   chimpanzee research?

                   JOHN PIPPIN
                   Senior Medical and Research Adviser
                   Physicians Committee for Responsible Medicine

                   JARROD BAILEY
                   Science Director
                   New England Anti-Vivisection Society

KEVIN KREGEL
Professor, Departments of Integrative
Physiology and Radiation Oncology
University of Iowa
Chair, Animal Issues Committee
Federation of American Societies for
Experimental Biology

4:15 p.m.   Discussion with the Committee

4:45 p.m.   ADJOURN

August 11, 2011

Keck Center, Room 100
500 Fifth Street, NW
Washington, DC 20001

Meeting Objectives:

- To obtain background data on the current use of chimpanzees in biomedical and behavioral research.
- To explore potential alternative models to chimpanzees.
- To seek public comment about the scientific need for chimpanzees in biomedical and behavioral research.

8:00 a.m.   Welcome and Meeting Objectives

JEFFREY KAHN, *Committee Chair*
Director and Professor
Maas Family Endowed Chair in Bioethics
Center for Bioethics
University of Minnesota

SESSION I: THE CHIMPANZEE

Session Objectives: Understand chimpanzee behavior and genetics and their role in biomedical research. Compare chimpanzees both to other models and to humans. Explore the usefulness of the chimpanzee as a model for biomedical and behavioral research, specifically for understanding human diseases and disorders. Discuss what scientific alternatives exist should the chimpanzee no longer be an available model.

JAY KAPLAN, *Session Chair*
Professor of Pathology (Comparative Medicine),
Translational Science and Anthropology
Wake Forest University Primate Center and
Wake Forest Translational Science Institute
Wake Forest School of Medicine

8:10 a.m.   Chimpanzee Behavior

FRANS DE WAAL
C.H. Candler Professor of Primate Behavior
Department of Psychology
Emory University

8:30 a.m.   Chimpanzee Genetics

JEFFREY ROGERS
Associate Professor
Department of Molecular and Human Genetics
Baylor College of Medicine

8:50 a.m.   Chimpanzee Biomedical Research

ROBERT PURCELL
Chief, Hepatitis Viruses Section
Laboratory of Infectious Diseases
National Institute of Allergy and Infectious
Diseases

9:10 a.m.   Panel Discussion with Committee

                •   What scientific alternatives exist should the chimpanzee no longer be an available model?

9:40 a.m.   BREAK

SESSION II: BEHAVIORAL RESEARCH

Session Objective: Review current use of chimpanzees for behavioral research. Explore alternative models also used in this research area.

> ROBERT SAPOLSKY, *Session Chair*
> Professor of Biology, Neurology and
>   Neurological Sciences
> Stanford University

9:50 a.m.   PANELISTS [15 min/talk]

        Chimpanzee Social Behavior and Communication

> WILLIAM HOPKINS
> Professor
> Department of Psychology
> Agnes Scott College

        Chimpanzee Learning and Memory

> CHARLES MENZEL
> Senior Research Scientist
> Language Research Center
> Georgia State University

        Potential for Non-Human Primates in Behavioral Research

> MARK MOSS
> Professor and Chair
> Department of Anatomy and Neurobiology
> Boston University

Chimpanzee Research in Zoos and Sanctuaries

> BRIAN HARE
> Assistant Professor
> Department of Evolutionary Anthropology
> Duke University

10:50 a.m.   Panel Discussion with Committee
  - What scientific alternatives exist should the chimpanzee no longer be an available model?
  - How long would it take for science to catch up if the chimpanzee were no longer available?

SESSION III: PUBLIC COMMENT

Session Objectives: Seek public comment from interested stakeholders about the continued and potential future need for chimpanzees in biomedical and behavioral research.

NOTE: To accommodate requests, speakers will be strictly limited to 3 minutes.

> JEFFREY KAHN, *Committee Chair*
> Director and Professor
> Maas Family Endowed Chair in Bioethics
>   Center for Bioethics
> University of Minnesota

11:20 a.m.   Public Comments

> ALICE RA'ANAN
> Director of Government Affairs and Science
>   Policy
> The American Physiological Society

> ANNE DESCHAMPS
> Science Policy Analyst
> Federation of American Societies for
>   Experimental Biology

JUSTIN GOODMAN
Associate Director
People for the Ethical Treatment of Animals

LAURA BONAR
Program Director
Animal Protection of New Mexico

STEPHEN ROSS
Assistant Director, Lester Fisher Center for the
    Study and Conservation of Apes
Lincoln Park Zoo

RAIJA BETTAUER
Bettauer BioMed Research

PAMELA OSENKOWSKI
Director of Science Programs
National Anti-Vivisection Society

SUE LEARY
President
Alternatives Research & Development
    Foundation

THEODORA CAPALDO
President/Executive Director
New England Anti-Vivisection Society/Project
    Release & Restitution

ERIC KLEIMAN
Research Director
In Defense of Animals

RYAN MERKLEY
Associate Director of Research Policy
Physicians Committee for Responsible Medicine

MATTHEW BAILEY
Vice President
National Association for Biomedical Research

JOSEPH ERWIN
Consulting Primatologist

KATHLEEN CONLEE
Director of Program Management
The Humane Society of the United States

BETH CATALDO
Director
Cetacean Society USA

CATHY LISS
President
Animal Welfare Institute

DAVID DEGRAZIA
Professor of Philosophy
George Washington University

C. JAMES MAHONEY
Research Professor
New York University School of Medicine

12:20 p.m. LUNCH

SPECIAL LECTURE

1:00 p.m. Chimpanzees in Biomedical and Behavioral Research

     JANE GOODALL *(via video conference)*
     Founder
     Jane Goodall Institute

1:30 p.m. Discussion with Committee

SESSION IV: HEPATITIS

Session Objectives: Review the role of chimpanzees in hepatitis research. Explore alternative models also used in this research area.

DIANE GRIFFIN, *Session Chair*
Professor and Chair
Department of Molecular Microbiology and
　Immunology
Johns Hopkins Bloomberg School of Public
　Health

1:40 p.m.　PANELISTS [15 min/talk]

The Current State of Hepatitis Research

ROBERT LANFORD
Scientist
Department of Virology and Immunology
Texas Biomedical Research Institute

The Next Drug for Hepatitis B and C

CHRISTOPHER WALKER
Professor of Pediatrics
Nationwide Children's Hospital
The Ohio State University

Cellular and Molecular Technique Advances in
　Hepatitis Research

STANLEY LEMON
Professor of Medicine
Division of Infectious Diseases
University of North Carolina School of Medicine

Humanized Mice for the Study of Human
Infectious Diseases

> ALEXANDER PLOSS
> Research Assistant Professor
> Laboratory of Virology and Infectious Disease
> The Rockefeller University

From Chimpanzee to Human—Translational Research
in Viral Hepatitis

> EUGENE SCHIFF
> Leonard Miller Professor of Medicine
> Director, Schiff Liver Institute/Center for Liver
> Disease
> University of Miami Medical School

2:55 p.m.　　Panel Discussion with Committee
- What scientific alternatives exist should the chimpanzee no longer be an available model?
- How long would it take for science to catch up if the chimpanzee were no longer available?

3:40 p.m.　　BREAK

SESSION V: INFECTIOUS DISEASES

Session Objectives: Review the role of chimpanzees in infectious disease research. Explore alternative models also used in this research area.

> JOHN BARTLETT, *Session Chair*
> Professor
> Department of Medicine
> Johns Hopkins University School of Medicine

4:00 p.m.　　PANELISTS [15 min/talk]

The Role of Chimpanzees in HIV Research

> NANCY HAIGWOOD
> Professor of Microbiology and Immunology
> Director
> Oregon National Primate Research Center

The Role of Chimpanzees in RSV Research

> PETER COLLINS
> Director
> RNA Viruses Section
> National Institute of Allergy and Infectious
>   Diseases

Current Experimental Models for Malaria Vaccine
    Development

> ANN-MARIE CRUZ
> Program Officer, Research and Development
> PATH Malaria Vaccine Initiative

Monoclonal Antibody Therapeutics

> THERESA REYNOLDS
> Director
> Safety Assessment
> Genentech

Alternative Models for Infectious Disease
    Research

> ROBERT HAMATAKE
> Director of HCV Biology
> GlaxoSmithKline

5:15 p.m.   Panel Discussion with Committee
- What scientific alternatives exist should the chimpanzee no longer be an available model?
- How long would it take for science to catch up if the chimpanzee were no longer available?

6:00 p.m.   ADJOURN

August 12, 2011

Keck Center, Room 100
500 Fifth Street, NW
Washington, DC 20001

Meeting Objectives:

- To obtain background data on the current use of chimpanzees in biomedical and behavioral research
- To explore potential alternative models to chimpanzees
- To seek public comment about the scientific need for chimpanzees in biomedical and behavioral research

SESSION VI: POTENTIAL FUTURE NEEDS

Session Objectives: Explore potential future needs for chimpanzees in biomedical and behavioral research. Consider emerging threats and novel technologies.

EDWARD HARLOW, *Session Chair*
Special Assistant to the Director
National Cancer Institute

8:30 a.m.   PANELISTS [15 min/talk]

Surveying the Future of Chimpanzee Research

THOMAS J. ROWELL
Director
New Iberia Research Center
University of Louisiana at Lafayettte

Is Chimpanzee Research Critical to the Health
Security of the United States?

JOSEPH BIELITZKI
Associate Director
Office of Research and Commercialization
University of Central Florida

The Role of Chimpanzees in Biodefense Research—
DoD Perspective

> JAMES SWEARENGEN
> Director (retired)
> Comparative Medicine Veterinarian
> National Biodefense Analysis and
>   Countermeasures Center

The Role of Chimpanzees in Biodefense Research—
NIH Perspective

> MICHAEL KURILLA
> Director
> Office of Biodefense Research Affairs
> National Institutes of Health

9:45 a.m.   Discussion with the Committee
- In the event of a public health emergency, what would the consequences be if there were no chimpanzees available for biomedical research?
- What would the impact be if chimpanzees were unavailable for testing during drug development and research?
- How long would it take for science to catch up if the chimpanzee were no longer available?

10:45 a.m.   ADJOURN

D

Committee Biographies

Jeffrey Kahn, Ph.D., M.P.H. (*Chair*), is the Robert Henry Levi and Ryda Hecht Levi Professor of Bioethics and Public Policy at the Johns Hopkins University Berman Institute of Bioethics. Prior to joining the faculty at Johns Hopkins in 2011, Dr. Kahn was director of the Center for Bioethics and the Maas Family Endowed Chair in Bioethics at the University of Minnesota, positions he held from 1996 to 2011. Earlier in his career, Dr. Kahn was director of the Graduate Program in Bioethics and assistant professor of bioethics at the Medical College of Wisconsin, and associate director of the White House Advisory Committee on Human Radiation Experiments. Dr. Kahn works in a variety of areas of bioethics, exploring the intersection of ethics and public health policy, including research ethics, ethics and genetics, and ethical issues in public health. He has served on numerous state and federal advisory panels, and speaks nationally and internationally on a range of bioethics topics. He has published more than 100 articles in the bioethics and medical literature, and is a coeditor of the widely used text *Contemporary Issues in Bioethics*, about to enter its eighth edition. From 1998 to 2002, he wrote the biweekly column "Ethics Matters" for CNN.com. Dr. Kahn earned his B.A. in Microbiology from the University of California, Los Angeles (UCLA), his M.P.H. from Johns Hopkins University, and his Ph.D. in Philosophy/Bioethics from Georgetown University.

John G. Bartlett, M.D., is an internationally renowned authority on AIDS and other infectious diseases. In 1970, he joined the faculty at UCLA. He later moved to the faculty of Tufts University School of Medicine, where he served as associate chief of staff for research at the Boston VA Hospital. In 1980 he moved to Baltimore as professor of medicine and chief of the Division of Infectious Diseases at Johns Hopkins University School of Medicine. For 27 years, he has been a leader for the School of Medicine's worldwide efforts to understand, prevent, and treat AIDS. He received the prestigious 2005 Maxwell Finland Award for scientific achievement from the National Foundation for Infectious Diseases. Dr. Bartlett was the first to direct clinical trials in Baltimore of new treatments that prevent HIV from replicating, and he pioneered the development of dedicated inpatient and outpatient medical care for HIV-infected patients. In 1984, when AIDS was still in its infancy, he helped start a small clinic within the Moore Clinic to serve a small group of gay men with AIDS, which along with providing research data about how the disease spread, grew to become the centerpiece of the Johns Hopkins AIDS Service. It is now the largest program for HIV care in Maryland. Dr. Bartlett cochaired the national committee that drafted the first and all subsequent treatment guidelines for HIV-infected patients. He counsels numerous medical societies and health ministries around the world on infectious diseases in general and on AIDS specifically. Bartlett's research interests have dealt with anaerobic infections, pathogenic mechanisms of *Bacteroides fragilis*, anaerobic pulmonary infections, and *Clostridium difficile*-associated colitis. Since joining Hopkins in 1980, his major interests have been HIV/AIDS, managed care of patients with HIV infection, pneumonia (community acquired), and, most recently, bioterrorism. Clinically his interests include HIV primary care, general infectious diseases, HIV and hemophilia, and HIV managed care. He received his undergraduate degree from Dartmouth University and earned his M.D. at Upstate Medical Center in Syracuse, New York. He then completed residency training in Internal Medicine at the Brigham and Women's Hospital in Boston and the University of Alabama at Birmingham. Dr. Bartlett also completed Fellowship training in Infectious Diseases at UCLA and at the Wadsworth Veterans Administration Hospital.

H. Russell Bernard, Ph.D., is the founder and current editor of the journal *Field Methods*, and has served as editor for the *American Anthropologist* and *Human Organization*. He has also served as the chair of the Board of Directors for the Human Relations Area Files. A member of the National Academy of Sciences (NAS), Dr. Bernard has been a recipient of the Franz Boas Award from the American Anthropological Association as well as the University of Florida Graduate Advisor/Mentoring Award. His teaching interests focus on research design and the systematic methods available for collecting and analyzing field data. He has taught both within the United States and in Greece, Japan, Germany, and England. Dr. Bernard received his B.A. in Anthropology/Sociology from Queens College, New York, his M.A. in Anthropological Linguistics from the University of Illinois, and his Ph.D. in Anthropology from the University of Illinois.

Floyd E. Bloom, M.D., is a past chair of the American Association for the Advancement of Science (AAAS), former editor in chief of the journal *Science*, and former chair of the Department of Neuropharmacology at the Scripps Research Institute in La Jolla, CA. A member of the National Academy of Sciences (NAS), he is the recipient of numerous prizes for his contributions to science, including the Janssen Award in the Basic Sciences, the Pasarow Award in Neuropsychiatry, and the Institute of Medicine's (IOM's) Rhoda and Bernard Sarnat International Prize in Mental Health. He has also been named a member of the Royal Swedish Academy of Sciences and a member of the IOM. Dr. Bloom's more than 600 publications include the seminal work, *The Biochemical Basis of Neuropharmacology* and *The Dana Guide to Brain Health*. In an important call-to-arms for healing the U.S. health care system, published June 13, 2003, in *Science* and based on his Presidential Lecture at the 2003 AAAS Annual Meeting, he describes how events of the 20th century have produced a system that cannot incorporate or implement new knowledge for the diagnosis or treatment of disease. Dr. Bloom earned his B.A. from Southern Methodist University and his M.D. from the Washington University School of Medicine.

Warner C. Greene, M.D., Ph.D., is director and Nick and Sue Hellmann Distinguished Professor of Translational Medicine of the Gladstone Institute of Virology and Immunology (GIVI), a research center that is affiliated with the University of California, San Francisco, and dedicated to fundamental studies of modern virology and immunology with a focus on HIV and AIDS. Dr. Greene graduated from Stanford University with a B.A. and Washington University School of Medicine with an M.D. and a Ph.D. He completed internship and residency training in medicine at the Massachusetts General Hospital. After serving as a senior investigator at the National Cancer Institute (NCI) and a Howard Hughes Medical Institute investigator and professor of medicine at Duke University, Dr. Greene moved to San Francisco in 1990 to become the founding director of the Gladstone Institute of Virology and Immunology. He is also a professor of medicine, microbiology, and immunology at UCSF. The ongoing research in Dr. Greene's laboratory focuses on the molecular basis for HIV pathogenesis, transmission, and latency and the biochemical mechanisms underlying the regulation and action of the NF-kB/Rel family of eukaryotic transcription factors. The lab also studies HIV Env-mediated fusion and its role in the transmission of HIV virions across the female genital mucosa. Dr. Greene's laboratory is also exploring how CD4 T cells die during HIV infection and devising new approaches to interdict this death pathway. Dr. Greene is the author of more than 330 scientific papers. He is a member of the IOM and a Fellow of the AAAS. He is also currently president-elect of the Association of American Physicians. In 2007 he became president of the Accordia Global Health Foundation, whose mission to build a healthy Africa where every individual can thrive. Accordia is specifically focused on overcoming the burden of infectious disease on the continent by creating innovative program models that strengthen health capacity, building centers of excellence, and strengthening medical institutions. With Paul Volberding, Dr. Greene also directs the UCSF–GIVI Center for AIDS Research and is a member of the executive committee of the AIDS Research Institute at UCSF.

Diane E. Griffin, M.D., Ph.D., has been the principal investigator on a variety of grants from the National Institutes of Health (NIH), the Bill & Melinda Gates Foundation, and the Dana Foundation. She is the author or coauthor of many scholarly papers and articles and is past president of the American Society for Virology, Association of Medical School Microbiology Chairs, and American Society for Microbiology. She is a member of the NAS, American Academy of Microbiology, and the IOM. Dr. Griffin began her career at Johns Hopkins as a postdoctoral fellow in Virology and Infectious Disease. After completing her postdoctoral work, she was named an assistant professor of Medicine and Neurology. Since then, she has held the positions of associate professor, professor, and now professor and chair. She has also served as an investigator at Howard Hughes Medical Institute. Dr. Griffin's research interests include alphaviruses, acute encephalitis, and measles. Alphaviruses are transmitted by mosquitoes and cause encephalitis in mammals and birds. She has identified determinants of virus virulence and mechanisms of noncytolytic clearance of virus from infected neurons. She is also working on the effect of measles virus infection and immune activation in response to infection on the immune system. In Zambia, she and her colleagues are examining the effect of HIV infection on measles and measles virus vaccination. They have discovered that measles suppresses HIV replication and are identifying the mechanism of this suppression. Vaccine studies are defining the basis for atypical measles, and a new vaccine to induce immunity in infants under the age of 6 months is under development using a rhesus macaque model. Dr. Griffin earned a Biology degree from Augustana College, followed by an M.D. and a Ph.D. from Stanford University. She completed her residency in Internal Medicine at Stanford University Hospital.

Edward Harlow, Ph.D., a distinguished molecular biologist, is an internationally recognized leader in cancer biology who is best known for his discoveries regarding the control of cell division and critical changes that allow cancer to develop. He is a professor of Biological Chemistry and Molecular Pharmacology at Harvard Medical School and a Special Assistant to the Director at the National Cancer Institute. Previously he served as Chief Scientific Officer of Constellation Pharmaceuticals, a Cambridge MA biotechnology company that specializes in making anti-cancer drugs that target the unusual transcriptional regulatory states found in tumor cells. He served as Scientific Director for the Massachusetts General Hospital Cancer Center and was Associate Director for Science Policy at the National Cancer Institute, where he helped direct U.S. cancer research planning. Dr. Harlow has received numerous scientific honors, including election to the National Academy of Sciences and the Institute of Medicine, appointment as Fellow of the American Academy of Arts and Sciences, and receipt of the American Cancer Society's highest award, the Medal of Honor. Dr. Harlow has served on a number of influential advisory groups, including the Board of Life Sciences for the National Research Council, External Advisory Boards for UCSF, Stanford, UCLA, and NYU Cancer Center, and Scientific Advisory Boards for the Foundation for Advanced Cancer Studies and numerous biotechnology and pharmaceutical companies, including Onyx, Alnylam, 3V Biosciences, and Pfizer Pharmaceuticals. He received his B.S. and M.S. from the University of Oklahoma and his Ph.D. at the Imperial Cancer Research Fund in London.

Jay R. Kaplan, Ph.D., is professor of Pathology (Comparative Medicine), Translational Science, and Anthropology at Wake Forest School of Medicine. He is also serves as head of the Section on Comparative Medicine (Department of Pathology) and director of the Wake Forest Primate

Center. He moved to the Wake Forest University School of Medicine in 1979 to study the effects of behavioral stress on susceptibility and resistance to coronary artery atherosclerosis in a monkey model of human heart disease. His current research with monkeys focuses on the behavioral and genetic factors that influence the quality of premenopausal ovarian function, and in turn, on the effect of ovarian function on risk for coronary heart disease and osteoporosis. This research has demonstrated that much of the postmenopausal trajectory for atherosclerosis and bone loss is established premenopausally, suggesting that primary prevention of postmenopausal disease should begin in the decades prior to menopause. His achievements include more than 150 peer-reviewed publications, the Irvine H. Page Arteriosclerosis Award for Young Investigators from the American Heart Association, the Presidency of the Academy of Behavioral Medicine Research, and the awarding of numerous grants from the NIH. He currently serves as principal investigator of the grant that supports a large pedigreed and genotyped colony of vervet monkeys. He also reviews for numerous journals and for the NIH. He has served as a member of the National Academies Institute for Laboratory Animal Research and as a member of the Animal Resources Review Committee of the National Center for Research Resources. Most recently, Dr. Kaplan became a member of the Society for Women's Health Research Interdisciplinary Studies in Sex Differences Fund for Cardiovascular Disease Network. He received his B.A. in economics from Swarthmore College. He then earned an M.A. and a Ph.D. in biological anthropology from Northwestern University, where his research involved behavioral observations of free-living rhesus monkeys on Cayo Santiago Island, Puerto Rico.

Margaret S. Landi, V.M.D., is vice president of Global Laboratory Animal Science (LAS) for Glaxo SmithKline Pharmaceuticals and Chief of Animal Welfare and Veterinary Medicine for GSK Research and Development. In this capacity, she is responsible for promoting animal welfare and providing a high standard of technical and professional assistance to the company's research and development community. Dr. Landi is a Diplomate in the American College of Laboratory Animal Medicine (ACLAM) and is a past president of the organization. Besides serving on the ACLAM Board of Directors, she has served on the Council of the Institute of Laboratory Animal Research (ILAR), a part of the National Academy of Science. While on the Council, she was editor-in-chief of the *ILAR Journal*. She serves currently on the Board of Trustees for the

Scientists Center for Animal Welfare, the National Association for Biomedical Research, and Americans for Medical Progress. Dr. Landi has received Distinguished Alumni Awards from both the University of Pennsylvania and William Paterson University. She has been awarded the Charles River Prize and the Pennsylvania Veterinary Medical Association's Veterinarian of the Year Award. In 2010, she was the recipient of the Harry Rowsell Award from the Scientists Center for Animal Welfare. Dr. Landi has published and presented papers on a number of topics related to laboratory animal medicine, welfare, and science. Her recent area of work is in the application of global principles for laboratory animals in an international arena where laws, cultures, regulations, and policies differ.

Frederick A. Murphy, D.V.M., Ph.D., is a professor in the Department of Pathology at the University of Texas Medical Branch (UTMB), Galveston. He is dean emeritus and distinguished professor emeritus of the School of Veterinary Medicine at the University of California, Davis (UCD). He is also distinguished professor emeritus at the School of Medicine, UCD. Earlier, he served as the director, National Center for Infectious Diseases, Centers for Disease Control and Prevention (CDC), and before that as director of the Division of Viral and Rickettsial Diseases at CDC. At UTMB, Dr. Murphy is a member of the Institute for Human Infections and Immunity (and its executive board), the Center for Biodefense and Emerging Infectious Diseases, the Galveston National Laboratory, the Center for Tropical Diseases, and the McLaughlin Endowment for Infection and Immunity (and member of its executive board). Dr. Murphy's professional interests include the virology, pathology, and epidemiology of highly pathogenic viruses/viral diseases: (1) Rabies: long-running studies leading to the identification of more than 25 viruses as members of the virus family *Rhabdoviridae*, identification and characterization of the first rabies-like viruses, and major studies of rabies pathogenesis in experimental animals, including the initial descriptions of infection events in salivary glands and in muscle; (2) Arboviruses: long-running studies of togaviruses and bunyaviruses with the initial proposal for the establishment and naming of the virus family Bunyaviridae, and characterization of "reo-like" viruses culminating in the establishment and naming of the virus genus Orbivirus; (3) Viral hemorrhagic fevers: long-running studies leading to the initial discovery of Marburg and Ebola viruses, and characterization of several other hemorrhagic fever viruses, culminating in the establishment and naming of the virus families

Arenaviridae (e.g., Lassa virus) and Filoviridae (Marburg and Ebola viruses), and elucidation of the pathology and pathogenesis of the diseases in humans, monkeys, hamsters, and guinea pigs caused by these exceptionally virulent agents; and (4) Viral encephalitides: long-running studies of the pathogenesis of neurotropic viruses in experimental animals, including alphaviruses, flaviviruses, bunyaviruses, enteroviruses, paramyxoviruses, herpesviruses, and others. He has been a leader in advancing the concept of "new and emerging infectious diseases" and "new and emerging zoonoses." Most recently his interests have included the threat posed by bioterrorism. Dr. Murphy has a B.S. in bacteriology and a D.V.M. from Cornell University, and a Ph.D. in comparative pathology from UCD.

Robert Sapolsky, Ph.D., is a professor of biology, neurology and neurological sciences, and neurosurgery at Stanford University. He has focused his research on issues of stress and neuronal degeneration, as well as on the possibilities of gene therapy strategies for protecting susceptible neurons from disease. Currently, he is working on gene transfer techniques to strengthen neurons against the disabling effects of glucocorticoids. Dr. Sapolsky also spends time annually in Kenya studying a population of wild baboons in order to identify the sources of stress in their environment, and the relationship between personality and patterns of stress-related disease in these animals. More specifically, he studies the cortisol levels between the alpha male and female and their subordinates to determine stress level. Dr. Sapolsky has received numerous honors and awards for his work, including the prestigious MacArthur Fellowship genius grant in 1987, an Alfred P. Sloan Fellowship, and the Klingenstein Fellowship in Neuroscience. He was also awarded the National Science Foundation Presidential Young Investigator Award and the Young Investigator of the Year Awards from the Society for Neuroscience, International Society for Psychoneuro-Endocrinology, and Biological Psychiatry Society. In 2007, he received the John P. McGovern Award for Behavioral Science, awarded by the AAAS. In 2008, he received the Wonderfest's Carl Sagan Prize for Science Popularization. Sapolsky received his B.A. in biological anthropology summa cum laude from Harvard University and his Ph.D. in neuroendocrinology from Rockefeller University.

Sharon F. Terry, M.A., is president and chief executive officer of the Genetic Alliance, a network transforming health by promoting openness as process and product, centered on the health of individuals, families, and communities. She is the founding executive director of PXE International, a research advocacy organization for the genetic condition pseudoxanthoma elasticum (PXE). Following the diagnosis of their two children with PXE in 1994, Ms. Terry, a former college chaplain, and her husband founded and built a dynamic organization that enables ethical research and policies and provides support and information to members and the public. Along with the other coinventors of the gene associated with PXE (ABCC6), she holds the patent for the invention, and with the assignment of all rights to PXE International, is its steward. She codirects a 33-lab research consortium and manages 52 offices worldwide for PXE International. Ms. Terry is also a cofounder of the Genetic Alliance Biobank. It is a centralized biological and data repository catalyzing translational genomic research on genetic diseases. The BioBank works in partnership with academic and industrial collaborators to develop novel diagnostics and therapeutics to better understand and treat these diseases. She is at the forefront of consumer participation in genetics research, services, and policy and serves as a member of many of the major governmental advisory committees on medical research, including the Health Information Technology Standards Committee for the Office of the National Coordinator for Health Information Technology, liaison to the Health and Human Services Secretary's Advisory Committee on Heritable Disorders and Genetic Diseases in Newborns and Children, and the National Advisory Council for Human Genome Research, National Human Genome Research Institute (NHGRI), NIH. She serves on the boards of GRAND Therapeutics Foundation, Center for Information & Study on Clinical Research Participation, The Biotechnology Institute, National Coalition of Health Professional Education in Genetics, and Coalition for 21st Century Medicine. She is on the steering committees of the Genetic Association Information Network of the NHGRI, the CETT program, and the EGAPP Stakeholders Group; the editorial boards of *Genetic Testing and Biomarkers* and *Biopreservation and Biobanking*, and the Google Health and Rosalind Franklin Society Advisory Boards. She is chair of the Coalition for Genetic Fairness, which was instrumental in the passage of the *Genetic Information Nondiscrimination Act*. She is a member of the IOM Roundtable on Translating Genomic-Based Research for Health. In 2005, she received an honorary doctorate from Iona College for her work in community engagement and haplotype mapping; in 2007 received the first Patient Service Award from the UNC Institute for Pharmacogenomics and Individualized Therapy; and in 2009 received the Research!America Distinguished Organization Advocacy Award. She has recently been named as an Ashoka Fellow and won the Clinical Research Forum's 2011 Public Advocacy Award.